CASE FILES OF THE ROCKY MOUNTAIN PARANORMAL RESEARCH SOCIETY

VOLUME 1

CASE FILES OF THE ROCKY MOUNTAIN PARANORMAL RESEARCH SOCIETY

VOLUME 1

Robert Lewis and Bryan Bonner

POLYMATH
—PRESS—

Aurora, CO

Case Files of the Rocky Mountain Paranormal Research Society Volume 1
by Robert Lewis and Bryan Bonner

This book contains the true case files of the Rocky Mountain Paranormal Research Society and is intended for educational purposes, to stimulate interest and conversation concerning the paranormal, history, and science. The views and opinions expressed herein are solely those of the authors and do not necessarily reflect those of any of Rocky Mountain Paranormal's clients. Every effort has been made to ensure accuracy of the information presented, but neither the Rocky Mountain Paranormal Research Society nor Polymath Press offer any guarantees nor will be held responsible for the use or misuse of any information contained herein.

Cover art: Bryan Bonner
Interior photographs: Bryan Bonner & Robert Lewis
Interior artwork: Aaron Bordner
Design and layout: Robert Lewis

Published by Polymath Press, a trade name of Polymath Enterprises, LLC. Please direct all inquiries to Polymath Press, P. O. Box 461870, Aurora, CO 80046-1870, online at www.polymathpress.com, or via email to editor@polymathpress.com.

First edition

ISBN (paperback): 978-1-961827-00-4
ISBN (eBook): 978-1-961827-01-1
Library of Congress Control Number: 2023944769

To all those whose lives are remembered through ghost stories and folklore.

We recognize the significance of the stories, and it is our desire to pay homage to your enduring presence. We extend our deepest gratitude for the privilege of retelling your tales.

Table of Contents

List of Photos and Illustrations

Introduction

The Rocky Mountain Paranormal Research Society (which throughout this book we'll also frequently refer to as either "Rocky Mountain Paranormal" or "RMP" in the interest of brevity) has been around for quite a while. While we're far from the first organization to investigate paranormal claims, we are nearly unique in our longevity. Most paranormal investigation groups (who might more accurately be called "ghost hunters" than "paranormal investigators," though that's a distinction we'll explain elsewhere) tend to pop in and out of existence faster than…well, faster than a ghost might flicker past the corner of your eye on a sleepless night.

There are a lot of reasons for our longevity, and we'll discuss how our philosophy lends itself both to more rigorous investigative technique as well as to that longevity in a later essay in this book. For now, the important thing to realize is that we've been around for approximately two and a half decades, and in that time, we've investigated countless places ranging from private residences belonging to people you've never heard of all the way to some of the biggest and most famous haunts in the world, and we've tackled claims ranging from ghosts to demons to aliens to monsters and everything in between. If it's weird, we're there.

The upshot of all that time and effort spent looking into bizarre claims is that we've seen a thing or two. Our case files are loaded with stories that deserve to be told.

Some of these stories are our own and some of them are what we've been able to uncover through both our detailed interview process and our deep dives into the historical literature. Some of the stories are extraordinarily well-documented (either by ourselves or by others) while others are mere stories, unverified (perhaps unverifiable), but nevertheless interesting.

There's an important lesson in that idea, as well as a narrow line we must constantly straddle. Just because there's a story, that doesn't necessarily mean it's true. Respect for truth and knowledge requires we hold our stories to the greatest degree of scrutiny possible. Conversely, however, just because a story is unverified doesn't mean it doesn't contain something of value. Even if a story is completely false, it may have something to teach us. The story "I had a cheeseburger and fries for dinner this evening and then went to sleep while watching a horror movie" may be a completely true story (it is), but it's so banal as to necessitate apologies for making you read it. On the other hand, Dostoyevsky's *Crime and Punishment* or Shakespeare's *Macbeth* may be completely fictitious, yet they contain that spark of humanity that makes them among the most valuable stories ever told.

Rocky Mountain Paranormal is staffed by people who really love a good story. We're old school, dyed-in-the-wool horror fans. If it's creepy, we love it. And so we see the value in a good ghost story regardless of its literal truth. But just because we see its value doesn't mean we accept its truth uncritically. Our society is also comprised of hardcore science fans, so we insist upon the most rigorous scientific study of any paranormal claim before we allow ourselves to accept it as true.

This all means we have a lot of different kinds of stories. In some cases, we've been able to conclusively explain the origin of the alleged paranormal occurrences. In other cases, we've been able to explain some of the things but not others. Some cases remain complete question marks in our ledgers. But they're all interesting in their own way. And that includes both the stories behind the paranormal claims and the stories of our investigations. Sometimes what happens while we're investigating ends up being at least as interesting as the alleged paranormal occurrence itself.

Over the years, we've spent at least as much time sharing stories as we have actually investigating claims. In addition to our research program, we've been dedicated (and hopefully entertaining) educators, bringing not only the ghost stories but the stories from history, the stories from science, and the stories from our personal experiences to audiences large and small. As exhausting as the lecture circuit can be, it's among the most rewarding things any of us have done.

But there's never been a complete central repository of all our stories gathered in one place and preserved for future researchers or curious minds to peruse. Yes, we've published reports from many of our cases on our website (www.rockymountainparanormal.com in case you want to pay us a visit), but the Internet seems more transitory and less tangible than a book. Plus, you'll notice we said "many" of our case files were archived on the website. It never quite contained all of them.

In late 2021, while discussing our plans for rebuilding our operation after a period of inactivity forced by the Covid-19 pandemic of 2020 and 2021, the concept of a book of case files seemed like an obvious step. We quickly realized, however, that a single book simply wouldn't suffice. This would have to be a multi-volume work. What you hold in your hands represents the first volume of that project. We make no promises regarding how many volumes this work will ultimately require because we're still out there investigating, so the case files remain a work in progress. Indeed, even some of the files documented in this book may be amended in future volumes as we revisit some of these locations in the future. But for now, these are the cases in their entirety, as they stand now. They represent decades of hard work, and we hope you find them as interesting as we do.

A word regarding the organization of these books seems to be in order. First of all, the order in which cases are presented in these volumes is not chronological. Our first thought was, in fact, a chronological presentation of our cases, starting with the earliest ones and working our way gradually to the present and beyond. The problem, though, is that there have been ebbs and flows in our activity over the years. A chronological presentation of these cases would necessarily require us to write clusters of truly fascinating chapters punctuated by long stretches in which our work was more humdrum. Furthermore, some of these cases represent years of ongoing research and their dates overlap substantially. We don't necessarily work one case to its conclusion before taking on another project. So would the chronology represent the date of our first investigation or our last? Clearly, the chronological approach wouldn't work.

We ultimately decided to arrange the case files into the books in such a way as to ensure a good mix of different "kinds" of cases in each volume. That is to say, we're mixing stories of ghosts with stories of aliens or UFOs. And each volume will contain some famous cases as well as some you've never heard about. Hopefully the result is that everything remains fresh and exciting throughout the reading process.

Within the volumes, we've divided the cases into three main sections. The first part contains our investigations of public venues (defined here as venues open to the public, not necessarily publicly owned) and cemeteries. These are the kinds of places you might be able to visit for yourself (though sometimes an appointment or specific business with the venue might be required). Part two contains our files from private residences or private individuals. These are not going to be the kinds of cases you've read about in the papers, but rather represent our attempts to help people understand the phenomena that have plagued them. The origin of the vast majority of these cases is a plea for help from a distressed and confused individual or family. Finally, the third part contains "other activities." That's a sort of catch-all term for everything else we do. A few involve our analyses of photos or footage published in the media. With few exceptions, these cases don't involve a lot of hands-on investigation, but we nevertheless use whatever techniques are necessary to explain the phenomenon in question. Beyond those, over the years, we've participated in a number of experiments and projects that don't quite fit into the other sections of these books but nevertheless merit inclusion in the case files. Those cases end up in that third section.

Within each section, the order of cases is not random but neither is it meant to suggest any overarching narrative. The order was chosen to provide a textured reading experience.

Surrounding the main three parts and interspersed throughout are a variety of essays. While not part of the case files themselves, these writings represent important portions of our guiding philosophy or lessons we've learned along the way. No doubt most readers are just here for the ghost stories, but don't neglect the essays; they provide a lot of useful context for the cases themselves.

It's our sincere hope that you find these stories educational and entertaining in equal measure. They represent not only decades of work but a true passion for everyone involved in our enterprise over the years. Some of them are creepy. Others may be humorous, moving, or just plain weird. Indeed, if you're anything like us, you'll often find yourself on an emotional rollercoaster as we transition from a tearjerker of a case to one that'll have you seeing red to another that'll leave you scratching your head.

So go ahead and read on to join us as we explore more than two decades of some of the weirdest experiences we've ever had.

The Rocky Mountain Paranormal Philosophy

There seem to be two major factions when it comes to paranormal claims. On the one hand, you have the believers, who tend to seek confirmation of the existence of the ghosts/aliens/entities in which they believe. On the other hand are the self-styled skeptics comprised primarily of non-believers who seek to debunk the paranormal claims. Rocky Mountain Paranormal proudly sits right between these two groups. When we happen to be in a contrarian sort of mood, we like to joke that we anger the believers and non-believers equally. In truth, though, we'd like to think we actually fit in rather well with both of these groups because our philosophy is compatible with what both sides claim to be doing.

One way to think about our overall philosophy is to distinguish between skepticism and what we can call cynicism. In this context, cynicism refers to rigid adherence to a foregone conclusion and not to Cynicism as in the ancient Greek philosophy pioneered by such thinkers as Antisthenes, Diogenes, and Crates of Thebes. In our sense of the word, "cynic" is often used as a derogatory term for those who disbelieve in whatever paranormal claim is being discussed. And sometimes the label applies. However, we maintain that there are cynics on both sides of the paranormal belief spectrum.

The cynical believer is one who absolutely believes in the paranormal claim in question and no amount of disconfirming evidence can change this individual's mind. Similarly, the cynical non-believer is one who absolutely does *not* believe in the paranormal claim and no amount of confirming evidence will ever be satisfactory. In the extremes, these individuals are rare. Most people, upon considering a certain weight of evidence, are capable of changing their minds. However, slightly less to the extremes of the spectrum exist a lot of people who may be *capable* of changing their minds but certainly don't seem very interested in giving alternative explanations a fair and open hearing.

True skepticism requires us to treat each case from a completely neutral point of view. That's not to say we need to disregard everything we know from prior experience, but we do need to consider the possibility that any given claim may or may not turn out to be an example of a genuine paranormal phenomenon (whatever that is— see the next essay for a definition).

An example seems to be in order, so let's consider a simple ghost story. Imagine Building X is supposed to be haunted by the spirit of someone who was murdered in the building some fifty years ago. The believer tends to want the story to be true, so even the slightest anomalous experience will be taken as proof positive that the entire story is true. The disbeliever tends to reject even the possibility of a genuine ghost or haunting and so considers even the slightest discontinuity in the story as proof positive that the whole thing is false. The true skeptic, on the other hand, takes a decidedly different approach and says: maybe the story is true, maybe the story is false, and we need to investigate (as impartially as humanly possible) to find out where the truth lies.

Rocky Mountain Paranormal insists upon the skeptical approach. We're open to the possibility of a ghost. We even hope that it's true. But we're going to look for any potential naturalistic explanations first. In our hypothetical example, one thing we'd certainly want to do is verify whether the alleged murder even took place. If we confirm it, that doesn't prove the ghost story, though if we debunk it, it certainly means at the very least that part of the story was wrong. We'll also examine Building X and try to account for every phenomenon people have witnessed therein.

And it goes even further than that. Imagine that, on one of our investigations, we're able to explain every portion of a claim through purely natural and mundane mechanisms. Does that mean ghosts don't exist? Hardly! It's always possible the next investigation could be a real ghost. Indeed, it doesn't even disprove the existence of the ghost of a murder victim at Building X. All it means is that, of the various phenomena once thought to be evidence of the ghost, they all turned out to be caused by something else.

The late great magician and skeptic James Randi once gave another great example. He imagined trying to investigate whether or not Santa Claus could use flying reindeer as a mode of transportation. As an experiment, he imagined taking a group of, say, 100 reindeer to the top of a skyscraper and, one by one, pushing them off. If one manages to fly, well, that's case closed. But what happens if, as seems so much more likely, they all instead fall to their rather messy deaths on the street below? Does that prove that reindeer cannot fly? It does not. It merely proves that our particular group of 100 reindeer couldn't fly *or chose not to.*[1]

Does that mean we're cursed to never know anything? If we're looking for absolute certainty, the answer is pretty close to "yes." Outside of pure mathematics, absolute certainty doesn't really exist. Scientists tend not to speak of "proof," but of evidence and the strength thereof. Even in courts of law, where we do use the word "proof," we only insist upon proof beyond a *reasonable* doubt, not beyond *all* doubt, because there's always room for some doubt. So what are we to make of something like the reindeer experiment? We can't claim absolute proof, particularly of a negative, but we can say that the outcome of the experiment seems to make it *less likely* that flying reindeer exist. We have to remain open to the possibility that they exist out there, somewhere. But if we keep experimenting often enough, eventually we keep moving that dial closer and closer to the "disbelief" side. Just as long as we never claim to move it all the way, we should be in good philosophical shape.

What about evidence in favor of something like a ghost? Imagine we watch an inanimate object move, seemingly of its own volition. Imagine further that we do our due diligence and rule out all the natural explanations we can think of (which might include hoaxes, a draft in the room, optical illusions, hallucinations, videographic anomalies, and the list goes on). Does that prove that a ghost exists and has moved the object? It does not. The more natural explanations we rule out, the more we might move our dial toward the "belief" side of the spectrum. But without the kind of absolute proof that simply doesn't seem to exist, we can never move it all the way.

An important thing to remember is that, just because we maintain that two possible interpretations are both possible, we don't necessarily have to consider them equally likely. This isn't going to turn into a lecture on inferential statistics (one of your authors has written such documents before and hopefully they'll be published soon), but that's essentially the whole point of statistical testing. We may not be able to remove all doubt, but we can at least use sophisticated mathematical techniques to quantify exactly how uncertain we are.

When it comes to paranormal investigation, there aren't a whole lot of statistical tests to be done. The kinds of evidence obtained during paranormal investigations are often qualitative rather than quantitative. We strive to obtain quantitative results as often as possible—measuring things is a key part of science—but the nature of paranormal claims frequently precludes the possibility. Nevertheless, even if it must be done qualitatively or informally, that gradual adjustment of the belief/disbelief ratio is how we gradually move toward truth.

Along the same lines, it's necessary to think not in terms of certainties, but in *degrees* of certainty. For example, we can be 100% certain of a mathematical theorem

---

1   Randi, J. (1992). *Pseudoscience and the Paranormal: Inaugural Skeptics Society Lecture* [DVD]. The Skeptics Society Distinguished Lecture Series.

that has been proved.[2] Similarly, we can each at least arguably be absolutely certain of our own existence (that's Descartes' *"cogito ergo sum,"* or "I think therefore I am"). We can be a little less certain, but still close, of the existence of things like objective reality and the physical objects that surround us. We can also be reasonably assured that the sun will rise and set on schedule tomorrow. And then we get to other things that are substantially less certain, but we nevertheless seem to take for granted: that we won't die in our sleep tonight, that our car will start tomorrow morning. And so on, all the way down to things we can be pretty certain *aren't* the case: flying reindeer seem like a good example, as would the idea that the Earth is flat. Which we would consider a paranormal claim, to be sure, but which we have thus far felt is beneath our dignity to investigate as one of our case files.

Paranormal claims, too, exist on a spectrum of plausibility. Extraterrestrial aliens make a great example. Given the immensity of the universe, many people are fairly comfortable with the idea that it seems likely for there to be some extraterrestrial life out there somewhere. It's much less likely that they've visited us on Earth. And is even less likely that they're shapeshifters that have secretly infiltrated our governments. That's not just to pick on those claims. It's a mathematical necessity that each successive claim in this chain is less likely than its predecessor. For aliens to have infiltrated our government, they must also have visited Earth. To have visited Earth, they must exist. By mathematical necessity, the probability of two things both being true is less than the probability of either one of them being true. This is especially easy to see when one claim is a subset of the other. Mathematical thinking like this, however, does not come naturally to the human mind. One requires substantial training to think in terms of mathematical logic.

Even when we don't have mathematical reason to say that one paranormal claim is more or less likely than another, there still seems to be that spectrum of plausibility. In our estimation, a cryptid like Bigfoot seems more likely to exist than a ghost (that's just our estimation—you're free to estimate otherwise). It's entirely possible that ghosts could exist and Bigfoot not, but given the spotty data, we estimate the best we can.

Regardless of those estimates, though, you will never find any of us claiming knowledge we do not possess. Even if we might think a particular claim seems unlikely to be true, we will do our best to approach our investigation from a position of neutrality. And to the extent that true neutrality is probably impossible for the human mind to achieve, we're very careful to build safeguards against bias directly into our methods, so we can at least maintain methodological neutrality even if our attempts at philosophical neutrality fail.

At some point, though, we need to make decisions about what to believe or what

---

2   At least within the constraints of our axiomatic system of mathematics— though if you really want to tumble down a mathematical rabbit hole, we suggest you should read up on Gödel's incompleteness theorems.

not to believe. Though we will always keep investigating as much as possible, when the time comes to reach a tentative conclusion, the best approach is to apply Ockham's Razor (also sometimes spelled Occam's Razor and also known as the principle of parsimony). This is a philosophical precept which tells us that, all else being equal, the simplest explanation is usually the correct one. This is often misunderstood, though. "Simplest," in this context, doesn't mean "easiest to understand." It means, instead, that the explanation we should consider more likely is the one that relies on the fewest unproven assumptions. Another way to think about it is that we should always try to choose the path from Idea A to Idea B with the fewest possible intermediate steps.

Because paranormal claims, by most definitions anyway, include the existence of unproved entities, Ockham's Razor tells us we should be reluctant to accept them. When mundane or naturalistic explanations exist, they are *more likely* to be the correct ones.

All of that having been said, and while keeping all of that skeptical philosophy firmly in place, there's another piece to the Rocky Mountain Paranormal philosophy that's a bit friendlier to the believers' side of things: we really *want* some of these claims to be true. A big part of the reason we look into paranormal claims is in the hope that someday we might be able to confirm one of them. We may not think it's particularly likely (again, depending on how "out there" the particular claim might be), but that doesn't mean we aren't hoping.

As we said before, we're horror fans through and through. There are few things in the world that bring us more joy than a good ghost story or tale of alien abduction. Even if we're somehow able to conclusively disprove one of the claims (as we have done in some but not all of the case files you're about to read), we still love the stories themselves. They're a part of our collective mythology—perhaps even part of what Jung referred to as the collective unconscious. And they're just plain fun.

The world is a big place, and undoubtedly there are still phenomena our science has yet to discover. Combine that with the number of people reporting paranormal experiences, and there's a strong temptation to think there must be something to all of this. When you add that to our predisposition to want these claims to be true, it would be very easy for us to abandon our skepticism and take even a whisper of evidence as proof of the claim. So we're constantly walking that tightrope between open-mindedness and skepticism.

It's probably true, as Shakespeare reminded us, that there are more things in heaven and Earth than are dreamt of in your, my, or anyone's philosophy. And with as many people as there are in the world, even if some paranormal event is exceedingly rare, there's a possibility that someone (perhaps a lot of different someones) may have experienced it. On the other hand, widespread belief in or experience of a claim does not necessarily mean it's true. Human senses are flawed and easy to deceive, and the human brain is remarkably bad at evaluating claims objectively. Paranormal investigation, properly done, is a balancing act between these competing ideas.

Beyond the question of the legitimacy of paranormal claims, though, is another piece of our philosophy: the storytelling piece. We take the position that, even if all the paranormal claims turn out to be false, they're still stories worth telling, because our stories tell us a lot about who we are. Not only are they fun, but they're often psychologically deep at a metaphorical or symbolic level.

And then there's the history. As the saying goes, those who don't learn from history are doomed to repeat it (unfortunately, someone once pointed out to us that those who do learn from history are often doomed to watch while everyone else repeats it). Alas, a lot of our history is being lost. Wonderful historic buildings are constantly being torn down, either due to disrepair or to make room for some new development. Some of the buildings you're about to read about are no longer standing, making these case files a kind of ghost story in and of themselves.

Perhaps even worse, people don't take the time to read the stories of the past. To many people, the study of history seems quaint or irrelevant. We maintain—no, we *insist*—this is not the case. Not only does our history provide the context for why things are the way they are today, not only does our history provide lessons for how we should proceed into the future, not only does history tell us about our own heritage, both individually and collectively, but the stories from history are the very voices of our ancestors. If you want to talk about ghosts or speaking with the dead, you really need to talk about history. Ghosts as literal entities may or may not exist—we continue to take our neutral stance—but the ghosts of the past can and do speak directly to us through our historic stories, documents, and buildings, if only we would take the time to listen to them.

Particularly when it comes to ghost stories, paranormal investigation offers a window not just into the fun and spooky tales of the ghosts themselves, but into our rich historical heritage. Part of our mission at Rocky Mountain Paranormal is to document and preserve that history to the greatest extent we can. As you'll learn about in Chapter 5, when a piece of history was almost lost, it was Rocky Mountain Paranormal, working in conjunction with a team of scientific experts, who managed to give a voice back to the dead in some small way. That's something of which we're incredibly proud, and it's something we try to do whenever we can. Both in the context of historical research and in the context of reminding people of history through our lectures (and now these books), we're using the paranormal stories as a framework to help preserve our history.

Given that paranormal stories are fun on their own and that they have such value in providing an excuse to talk about history and science, it's little wonder that some of us have found such passion in this field.

And that passion is directly related to our guiding philosophy. To sum it up as succinctly as possible, our philosophy is that there's great value in paranormal stories, but they must be evaluated skeptically and scientifically to determine what is (or what is likely to be) true.

Or to put it another way: our guiding philosophy rests on equally important pillars of history, horror, storytelling, science, and skepticism.

A Brief Note on Methodology

There's no way we can provide a complete course in paranormal investigation here in this book.[3] However, it's worth taking a moment to briefly explain some of our methods, so you'll understand our approach in the case files to follow. As we discuss this method, you should be aware that the precise order of events is variable, and everything always depends on the specific paranormal claim being made. But just to provide a sketch of a "typical" investigation, we're going to walk you through the steps we would ordinarily take.

In the case of a private residence, the process usually starts with a client reaching out to us, usually via email. Hopefully we're the first team they reach out to, because sometimes other teams have confused or frightened the clients even more by the time the case reaches our desk.

Regardless of whether we're the first contact or not, we open a case and request further information from the client. Typically the claims made in these initial letters are vague enough that we don't know whether we'll be able to help or not. So we reach out to the client and request additional details. As things progress, if the case sounds like something with which we may be able to assist, we'll schedule an interview.

During the initial contact and interview phases, we try to collect detailed information regarding what is happening and when it's happening. We almost always request the client should keep a diary of occurrences, so we can check for any patterns. If the phenomenon always occurs on Wednesday evenings at 10:00 p.m., for instance, it wouldn't do for us to try to investigate at noon on a Saturday.

We also try to collect as much information about the clients themselves as they're willing to share. This includes any potential substance abuse problems, psychological issues, or major life changes. Additionally, we often counsel people to speak with a licensed psychological professional. This is *not* because we're calling them crazy. Though the possibility exists that people's paranormal issues could stem from deep psychological troubles, the more common reason we need to ask those questions and provide that advice is that paranormal events (whether we assume them to be genuine or psychological in nature) tend to occur during times of profound stress. Furthermore, the experience of paranormal phenomena itself can prove stressful or strain family relations, and it's important people should seek the help they need.

Once we have a detailed account of the claimed phenomena, including a diary or timeline, and possibly even including photos or videos the clients may send us,

3   Stay tuned: we're planning such a technical book at some point in the future.

we schedule a time for an on-site investigation if we deem it beneficial. The scheduling can be difficult because we need to coordinate times that work for our team as well as for the clients, but most importantly which correspond with the timing of the phenomenon in question. Though it can be difficult, we absolutely will rearrange our schedules for these investigations when necessary.

Before we arrive at the site (and often again after leaving the site), we go through an extensive process of background research. This includes digging through newspaper archives, public records including real estate transactions, and genealogical records if they seem relevant. We spend a lot of time on the Internet, at the state archives, or at the libraries[4] finding any information that might be relevant to the case. If possible, we try to obtain blueprints of the site. If these are unavailable, we'll make our own approximation of a site map during our investigation.

When we arrive at the site, we're not doing what you see on television. First of all, we don't bring a television crew (much to the surprise of some of our clients). We also ask them not to invite guests. Though we realize our work is interesting to people, we need the site to be as controlled as possible, and that means not having a bunch of people running around contaminating data. We'll then tour the building, determine locations for cameras and other monitoring equipment, set up our base camp, and prepare for the investigation. We ask the clients to pretend we're not there to the greatest extent possible and continue with their normal routine—it is, after all, during their normal routine that the phenomenon is claimed to occur.

Most importantly, once everything is set up, we shut up. We don't try to carry on conversations with the ghosts or run around talking to each other. Noise contaminates data. Furthermore, if we start talking to the alleged ghosts, we'd be psychologically priming ourselves not only that there is a ghost (which is what we're trying to determine in the first place) but also to expect specific answers to our questions. We might announce something like "if there are any spirits, feel free to make yourselves known to us" (with heavy emphasis on the word "if,") but that would be the extent of it. If the particular claim necessitates a more direct interaction with the alleged entity, we are careful to always do so hypothetically.

Equipment we might use on site includes still cameras, video cameras, 360-degree or spherical cameras, electromagnetic field (EMF) detectors, seismometers, thermometers, audio recorders, and any other tools that might seem relevant to a specific case. We sometimes also include air quality monitors to alert us if the house may have something like a gas leak (which can, under some circumstances, cause hallucinations).

Because the paranormal is an ill-defined field of study, we don't know how to

---

4   To anyone interested in any kind of research (paranormal or otherwise), take note: librarians are your best friends. They don't just spend their time rearranging the bookshelves; they're almost like wizards in their ability to find whatever information you need.

measure a ghost, alien, or demon. What we can do is establish baseline readings on our various instruments and then look for anomalous readings. If an anomaly should occur, we don't immediately assume it's supernatural. Instead, we note it, and follow up to see if we can find a way to explain it.

There are plenty of devices and gizmos claimed to be part of the ghost hunter's arsenal that we tend not to use most of the time. These are the devices that were specifically designed for paranormal investigation. Such devices are either hoaxes or intended for entertainment purposes rather than forensic investigation. There's no way to calibrate a device to detect a ghost since we've never even conclusively proved that ghosts exist or figured out their properties. We do own these devices and will experiment on or with them, but they're not part the gear we'll use on site unless the claim in question is specifically related to one of them.

At the end, we'll present the client with a report of our findings. This will include all of our background research as well as all of our measurements. The most important part will be our lists of findings, and they come in three categories. First, we list explainable events, or apparently paranormal phenomena we were able to explain naturally. Next, we list any phenomena or events we witnessed but have not yet been able to explain. Finally, we list any phenomena the clients reported but we were unable to witness or recreate. Along with this final list, we'll include any possible explanations we can think of, with the caveat that until we actually witness or test it, those explanations are speculative.

After we present our report, the clients often still have difficulty dealing with their issues. We're happy to help them to the extent we're able, but we always encourage them to speak to a psychological professional or work with their own clergy.

You'll note that our methods differ from what you tend to see on television in which the so-called investigators run around in the dark scaring themselves and each other. We mark a distinction between the activities of "ghost hunting" and "paranormal investigation." Ghost hunting is what you see on TV. It's a lot of fun, but it doesn't yield quality data. Paranormal investigation is methodical, scientific, and often quite boring or tedious to do.

Ghost hunting does have its place. It actually has two purposes. The first is purely for entertainment. As horror fans, we can appreciate running around in the dark and scaring ourselves silly. We just don't claim that to be part of a scientific investigation, no matter how many beeping gizmos we carry with us when we do so. The second is that ghost hunting can actually precede a proper scientific investigation and can be valuable in that context. It gives the phenomenon a chance to manifest itself under less stringent controls. The important thing to realize is that, when those controls are loosened, everything observed must be taken with a grain of salt.

In mathematics, there are two kinds of data analysis: exploratory and confirmatory. The former is essentially mining data to see what we can find. The latter is to confirm the validity or truthfulness of whatever we found in the former. Ghost hunting, in its proper context, can be thought of as a kind of exploratory analysis. Go to a haunted place and see what happens. But, as is the case in mathematics, you *must* follow it up with more rigorous confirmatory analysis.

ESSAY
Defining the Paranormal

Since we are in the business of paranormal research and investigation, one of the things we ought to be able to do is to provide a definition of what the paranormal actually is. Remarkably, it seems like relatively few people in this field take the time to give deep thought to the very nature of what they're doing. Exactly what is the paranormal? And how might one go about investigating it scientifically?

A good place to start is clearing up a rather subtle distinction between two related terms: paranormal and supernatural. Whenever someone uses either of these words, they tend to be thinking of the same kinds of claims or entities: ghosts, demons, spirits, etc. But there are important distinctions between the two. Etymologically speaking, the word supernatural comes from the Latin "super," meaning above, beyond, or outside of, and "natura," referring to nature. On the other hand, the word paranormal comes from the prefix "para," which means "alongside," and the word "normal." Therefore, supernatural claims or entities are ones thought to be beyond or outside of nature. Paranormal claims, by contrast, are simply beyond or outside the norm.

As we said, it's a subtle distinction. For one thing, any entity that exists outside of nature almost by definition also exists outside of the normal. But the converse is not necessarily true. We can imagine all kinds of things that are outside of the sphere of normal experience but nevertheless fall under the category of "natural."

Some examples will illustrate the point. In most religions, God is thought to exist outside of the natural universe. By definition, this makes the deity a supernatural entity. Cryptid creatures, by contrast (think of Bigfoot or the Loch Ness Monster as stereotypical examples) aren't necessarily thought to be supernatural. Most people seem to think of them as animal species just like any other except perhaps in their size and rarity. Because encounters with such animals is outside of normal experience, we have no problem classifying them as paranormal but not supernatural.

There are more difficult cases. Consider a ghost. If it exists, is it supernatural? Most concepts of what a ghost is would say so. The soul or spirit is thought to be more akin to a religious entity and therefore outside of natural law. The natural body, following natural law, may die, but the ghost somehow lives on supernaturally. However, it's not necessary to think in those terms. To be certain, ghosts are unknown to today's natural science, but if we were to discover conclusively that they exist, it's possible to think they're not outside of nature but that they represent an as yet undiscovered mechanism within natural law. An analogy might be quantum mechanics. The results physicists find in quantum mechanics are certainly bizarre, but the fact that they were discovered only recently doesn't mean that they were any less a part of nature prior to their discovery. If we proved the existence of ghosts tomorrow, there's a fair chance that the discovery would be accompanied or followed by the discovery of natural mechanisms that allow such an entity to exist.

Our operating definition of the paranormal encompasses anything that might be considered supernatural *as well as* anything that might be considered natural but is as yet unknown. It's a bit of a self-serving definition, admittedly, because the breadth of our interest extends to anything and everything that might seem weird, creepy, or bizarre. We therefore adopted a definition that encompasses just about anything and everything we might want to look into: things that are not (yet) known to science, or which science is only just beginning to understand.

There's a doctrine of naturalism within the philosophy of science. The argument is that science, as a method for studying the natural world, can only study natural phenomena. Supernatural entities may not be disproved by science, but neither are they scientific ideas or subject to scientific scrutiny. Science and the supernatural exist in what the late paleontologist Stephen Jay Gould called "non-overlapping magisteria."[1] We don't necessarily agree with that philosophy, at least when taken to its most restrictive extremes.

We take the position that paranormal (even potentially supernatural) phenomena can be and must be subject to scientific scrutiny to whatever extent they interact with the natural world. If we conceptualize a ghost as an entity that exists completely in a supernatural realm and never interacts with the living, there's no way to investigate it. In that case, we take the story for what it's worth, but our work is done before it even begins. That, however, isn't what people tend to claim. They claim to see ghosts, to feel ghosts, or to experience other effects of ghostly behavior. Because those phenomena

---

1   Gould, S. J. (1997). Nonoverlapping Magisteria. *Natural History, 106:* 16-22 & 60-62.

exist within (or at least affect) the natural world, science can examine them, at least indirectly.

That's where scientifically-minded groups like the Rocky Mountain Paranormal Research Society come in. We're not practicing scientists operating in a formal institutional laboratory somewhere, though we are scientists in the true sense of the word (people who perform scientific research), and most of our members have some kind of scientific background or training. The simple truth of the matter is: most scientists don't have the time to work on claims at the fringes of science. Career advancement in science is dependent largely on one's ability to perform publishable research, and in order to get published, one's research usually needs to yield positive results (negative results and inconclusive results seldom receive attention from the highly respected scientific journals). Even if they might be interested in "tumbling down the rabbit hole," so to speak, and working on some fringe ideas just on the off chance they might turn out to be true, it's simply not practical for most professional scientists. Our job is to pick up that slack and work on those fringe claims that are often overlooked, ignored, maybe even shunned by mainstream science.

With all of that out of the way, we return to our question at the beginning. Just what is the paranormal? For us, it can simply be defined as any entity, practice, or phenomenon outside of the realm of normal experience and/or scientific understanding. Typical examples include but are by no means limited to: ghosts, demons, aliens/UFOs, monsters/cryptids, psychic abilities, astrology, and many, many more.

PART ONE:
Public Venues and Cemeteries

While most paranormal claims are experienced and dealt with privately and seldom make the newspapers (and even if they do, tend to be quickly forgotten), there are some places that have become famous for their alleged paranormal phenomena. Many hotels and former hospitals are thought to be haunted. Same goes for plenty of cemeteries. In this section's fourteen chapters, we're going to look at some of Colorado's most famous haunts as well as some lesser-known but still fascinating settings for alleged paranormal phenomena.

ESSAY
The Profitability of a Haunted Business

Imagine you own a business. Say it's one that's open to the public. Maybe a hotel, possibly a bar or restaurant. Something of the sort. Imagine further that either you've begun to experience paranormal phenomena yourself or that your business has a reputation for being haunted. What are you to do? Is it better to advertise the haunting? Try to shut it up? Sell the business and move?

The answer largely depends on the business owner's own predilections. Religious people who come to the conclusion that their place of work is possessed by a demon or other evil spirit aren't likely to want to stick around. On the other hand, paranormal enthusiasts who think their building is haunted by a ghost are likely to not only stay put but to advertise it to their fellow paranormal enthusiasts. The latter are frequently (but not always) the ones that end up getting famous, because their owners invest time and money into advertising the presence of ghosts as an attraction to curiosity seekers from all around the world. The question, though, is whether that's an effective marketing strategy.

Most real estate agents will tell you that a house with a reputation for being haunted is typically slower to sell (and for a lesser price) than an equivalent property without such a reputation. The same is true for houses that have been the site of murders or other notorious crimes. Sometimes, this makes sense in a purely naturalistic way; a former meth house is likely not as valuable due to the possibility of health hazards or structural damage. But in the case of a haunted house, it seems to mostly be people's psychological unwillingness to stay in such a house that makes it lose its value.

These are called "stigmatized properties," because they have some kind of repu- tation or history that makes them potentially undesirable for buyers. Stigmas include crimes committed at the house, debts, and, yes, allegations of alleged paranormal phe- nomena. In the United States, state law governs what (if anything) sellers or real estate agents must disclose regarding stigmatized properties prior to a sale.

Of course, those laws vary from state to state. Most states require "physical defects" to be disclosed. These are potential defects that may affect the safety, habit- ability, or structural soundness of the property. But some states also require disclosure of "emotional defects," which may include criminal histories or paranormal claims.

Several states have laws concerning the disclosure of deaths that may have oc- curred on the property. Alaska requires disclosure of any murder or suicide on the property in the year preceding the sale, for example.[1] Conversely, Arkansas statute spe- cifically relieves sellers of the burden to disclose a murder or suicide on the property.[2] It's quite strange how widely the laws vary.

Interestingly, four states specifically mention allegations of paranormal activity in their laws. These merit individual mention. According to Massachusetts law, sellers do not need to disclose any psychological stigmas, including potential paranormal activi- ty.[3] The same is true in Minnesota: sellers don't need to disclose facts that may affect a buyer's psychological enjoyment of the property, including "perceived paranormal activity."[4] In New Jersey, sellers are not required to volunteer these kind of facts, in- cluding related to paranormal activity, but must disclose them if the buyer specifically asks about them.[5] New York law is a bit of an odd man out. According to statutory law, there is no duty to disclose[6], except as established by court precedent. According to that precedent, established in the 1991 New York Supreme Court ruling in *Stam- bovsky v Ackley*[7] (commonly known as the Ghostbusters ruling), the court will rescind a sale if the seller himself or herself created or publicized the reputation for paranormal activity (such as by holding haunted tours of the property) and takes unfair advantage of the buyer's ignorance of that reputation.

Since we're based in Colorado, we should also note that Colorado is one of the states that does not require disclosure of psychologically stigmatizing facts such as deaths or murders within the property.[8] While state law does not specifically address

---

1   Alaska Statutes 08.88.615 c. 1-2.

2   Arkansas Code 17-10-101.

3   Massachusetts Law, Part 1, Title XV, Chapter 93, Section 114.

4   Minnesota Statute 513.56.

5   New Jersey Administrative Code 11:5-6.4(d).

6   Consolidated Laws of New York, Real Property (RPP), Chapter 50, Article 12-A, Section 443.

7   169 A.D.2d 254 (N.Y. App. Div. 1991).

8   C.R.S. 38-35.5-101.

paranormal claims, the most likely interpretation given the statute concerning disclosure of deaths is that disclosure of paranormal activity or reputation is not required in the state of Colorado.

But the fact of the matter is, regardless of disclosure laws, buyers can often find out whether a property has a stigma attached to it through other means, and this can affect the value of real property. Many people have no interest in living in a haunted house, and real estate agents have told us that a reputation for being haunted can significantly decrease the price for which a house might sell. Though it seems to us that there's enough interest in the paranormal that it could be a selling point for the right buyer.

But that's residential property. The same phenomenon doesn't seem to affect commercial businesses. Indeed, in the case of commercial properties, the opposite seems to be true: they *want* the ghost stories. Of course there are business owners who want nothing to do with ghost hunters or the paranormal, but many businesses find a way to turn their ghost stories into profit, using their buildings' reputations as a means to attract curiosity seekers. They might hold ghost talks[9] or even just use the paranormal as an excuse to get some free advertising by getting their stories written up in the local newspapers.

Many of the businesses we've investigated over the years have fully capitalized on their haunted reputations. Probably the most famous one in Colorado is the Stanley Hotel in Estes Park (which we'll get to in Chapter 14). In fact, they even host theatrical seances performed by magicians every night, daily haunted tours, and more, right on their own property. It seems like the bulk of their business these days comes not from being a magnificent old hotel with spectacular access to the Rocky Mountains (which they are and they have), but from their ghostly reputation. Other businesses have, to a greater or lesser extent, followed the same business model.

It's important to note that, when we investigate a supposedly haunted business, we're not going to tailor our results to fit with their marketing. We're going to report whatever we find as truthfully as we can. Occasionally this has caused friction between our investigators and business owners who didn't like that we had managed to discredit some portion of their ghost stories. What they didn't realize, but we hope they will come to realize, is that we like the ghost stories just as much as they do. We're not trying to cut into their profits. In fact, we're going to advertise them even more by repeating those ghost stories. We're just *also* going to repeat the results of our investigation. We think they just add to the rich tapestry that is paranormal lore.

---

9   We're available for hire.

1
The Haunted Newspaper: The Colorado Springs Independent

The Colorado Springs Independent (also known as the Colorado Springs Indy or just The Indy) is a weekly newspaper serving Colorado Springs and the Pikes Peak region. It's Colorado's largest locally owned media company. *The Denver Post*, by contrast, though a substantially larger newspaper in terms of both staff size and circulation, is not entirely under local control but is owned by a Manhattan-based hedge fund. The Indy also publishes the *IndyBlast* website, the *Colorado Springs Business Journal*, and a variety of smaller niche publications.

But it's not their news services that caught our attention. Their headquarters, located at 235 S. Nevada Ave. in Colorado Springs, has only been the property of The Indy since their founding in 1993. But the property's more than a century of history began with a church. And of course, as is the case with many of these wonderful historic buildings, it's home to plenty of ghost stories.

Figure 1.1. The Colorado Springs Independent. Photo: Bryan Bonner.

The History

The property eventually destined to become Colorado's largest locally owned newspaper began its life in the hands of one Reverend R. C. Bristol, a local pastor. Rev. Bristol's name isn't one you often encounter in local histories, but it seems he was the head of a small local congregation and by all accounts had great fondness for the Colorado Springs region. Indeed, he was present at a meeting that determined Colorado Springs should be the home of Colorado College[1], and went on to serve as a trustee of that newly founded institution. Rev. Bristol owned the property (then just an empty lot) from 1885 through 1889, and during that time left the property to his wife Ruth Bristol.

Ruth Bristol, in turn, gave the property to the Church Erection Society in 1890.

Meanwhile, in 1887, a church known as the Tourist Memorial Mission Church was founded in Colorado Springs under the ministry of Revered W. E. McCormick. Though not officially chartered until the mid-1890s, most of the congregation were members of the United Brethren in Christ, a protestant denomination with connections to the Episcopals and the Methodists. When Revered J. Oliver became the pastor in 1895, the Church was officially chartered and its name changed to Tourist Memorial United Brethren in Christ Church. At that time it was located on Las Vegas Street in Colorado Springs, where it remained for a period of some years before relocating to a property at 417 S. Cascade which was later sold to the Sons of Israel. In 1911, Revered Henry Kohler became the pastor and presided over the purchase of the Nevada Street property for the sum of $6,000 (around $187,000 in 2022).

At this time, the congregation consisted of around 200 members, but the property was still an empty lot. Congregants would celebrate their services in a large tent while a basement for the building was being dug. This basement was completed in <u>1912, but construction</u> of the upper portion of the building would not be finished

---

1   Though it's a bit beyond the scope of our discussion of the ghost stories at the Colorado Springs Independent, we'd be remiss if we didn't mention some other interesting local history here. Colorado College was founded in 1874 by Reverend Thomas Nelson Haskell (with assistance from Rev. Bristol and others) in his daughter's memory, and in the years since has gone on to become one of the most respected private colleges in the United States. It was during her summer teaching position at Colorado College in 1893 that Katharine Lee Bates wrote a poem originally entitled "Pikes Peak," which would later, paired with music composed in 1882 (and published in 1892) by Samuel A. Ward, become the famous patriotic song "America the Beautiful." That rendition, combining Bates' words with Ward's music, appeared in 1910. Bates and Ward never met. Unfortunately, Ward died in 1903, never knowing the fame his music would achieve, though Bates was fortunate enough to see some of the popularity of the song before her death in 1929.

until 1917. To this day, a cornerstone on the southwest corner of the building reads: "United Brethren Church 1912 and 1917."

Beginning in 1911 and continuing until 1953, the Ladies Aid Society of United Brethren hosted a weekly "Chicken Pot Pie Dinner" on Thursdays to fund the church's activities. Unfortunately, several times strong winds ripped or otherwise damaged the large tent structure. The congregation's women responded by bringing their own sewing kits to repair the tent in time for Sunday services.

In 1915, trouble erupted and the congregation split, leaving only 27 families in the congregation under the supervision of one Reverend R. J. Parret, who was succeeded by Reverend I. A. Chivington, who served until 1916, when Reverend W. G. Schaefer, a freshly minted reverend from Bonebrake Seminary, took over the church, changing its name to First United Brethren of Christ Church.

At this time, the church still had no proper building due to a confluence of factors: the small congregation, splits within the church community, and perhaps most significant of all, the economic toll of World War I. As such, the locals began to call it the "Hole in the Ground Church."

But Rev. Schaefer, despite his relative inexperience, turned out to be the right man for the job. He reached out to the congregation and local community with an innovative "Buy a Brick" program to support construction of the church. The community responded. Following the sale of 200,000 bricks through Rev. Schaefer's program and aided by a loan from the Church Erection Society in Dayton, Ohio, the church was finally finished in the Late Victorian/Romanesque Revival style, along with a five-room parsonage, in 1917.

In 1946, the Evangelical and United Brethren denominations merged. The church changed its name to First Evangelical United Brethren Church.

In 1950, the Church built an educational wing on the East side of the church building known as the Schaefer Christion Educational Unit. The addition was designed by architect Grant A. Wilson and construction was completed by contractors George O. Teats and Sons. The addition was dedicated in May of 1952.

In 1957, Rev. Schaefer finally retired after 41 years of service and passed the reins to Reverend George L. and Mrs. Helen Edie, with Reverend Bruce Grauberger joining as assistant pastor in 1958. The main sanctuary underwent remodeling in 1960. In 1963, the Edies were relocated to a congregation in Aurora, Colorado and Reverend Paul Gamber became the new pastor.

In 1968, the church again underwent a name change. The Evangelical United Brethren and Methodist denominations merged and the name changed briefly to the United Methodist Church. However, because First United Methodist Church of Colorado Springs had already been established, the Nevada Avenue church changed its name again in the very same year to Central United Methodist Church.

By 1970, the congregation was beginning to shrink as many members of the community were relocating to the northeast portion of Colorado Springs. When Rev. Gamber retired that year, Reverend Tom Bennauzar was appointed the new pastor. The congregation and church leadership began discussing a move of location, likely due to the declining membership.

In 1971, Reverend Keith Spahr was made the head of the church, and discussions of relocation became more serious. A "for sale" sign was placed out front and the building was sold in 1973 to the Colorado Springs Police Department to be used as a police training academy. In addition to training new recruits to the police service, the building also housed a variety of local youth programs and the District Attorney's juvenile diversion program.

Following the Police Department's stewardship of the building, in 1992 it passed through the hands of the Smokebrush Foundation for the Arts, becoming the Smokebrush Center for the Arts, and remodeled to include art studios, galleries, and a 200-seat theatre. The arts center closed in 2002, and the building was purchased in 2003 by Codependent, LLC, the parent company of the *Colorado Springs Independent*, in whose possession it remains to this day.

Paranormal Claims

The Indy has its share of ghost stories. Part of its reputation for being haunted may be due to its history as an old church building. Old buildings in general and old churches in particular seem to attract ghostly or paranormal lore. Partly it's just about the age of the buildings and the amount of history to which they've borne witness. Likely it's also because churches are typically designed with architecture intended to inspire awe. When combined with age, that awe can easily turn into a sense of general creepiness.

The following ghost stories were told to us by members of the newspaper's staff during our interview process. Names have been redacted to protect the witnesses' identities (initials used to identify each witness do not necessarily reflect their actual initials).

J—'s Story

J— was a newspaper employee just starting her first day of work. While tending to her affairs inside the building, she had opportunity to glance out a window overlooking the front steps of the building. What she saw was haunting and terrifying: a woman with long red hair being dragged down the steps leading from the building to the sidewalk.

The front door of the building is marked by a porch with two columns at the head of the staircase in question. At the foot end, the staircase breaks into two smaller stairways leading to the sidewalk.

J— immediately rushed to the poor woman's aid, but when she opened the doors to investigate, nobody was there. No further explanation for this event has ever been found.

On another occasion, J— reports seeing a woman in the lobby. This woman had dark brown hair and a wore a black and white dress and a hat, in what J— describes as a 1950s or 1960s style. J— approached the woman to ask if she could help her, but of course the woman wasn't really there—or at least not among the living. According to J—, this woman has made several appearances in the building.

Another apparition J— describes is a man dressed in clergyman's garb. J— says she's seen this man three or four times, typically either by the kitchen or the "morgue," the newspaper's name for a room used to store papers (not for storing bodies). This clergyman is supposed to be around six feet tall, with brown hair, and apparently in his mid-60s. When he appears, he looks like he's preparing for a talk or a sermon, fixing his collar in the mirror or shuffling papers. On one occasion, he walked directly toward J— but gave no impression that he'd seen her.

Figure 1.2. The "morgue" storage area. Photo: Bryan Bonner.

On multiple occasions, but always on a Monday, J— claims to have heard a little girl saying "help me" in the kitchen or restroom. She's also seen a small child run around the front counter, only to vanish. This occurred when no children were known to be present in the building. It's unclear whether these are different children or whether they're one and the same.

Often, while working, J— felt a breeze or draft in her office even though the windows and doors were shut.

Once while filling papers in the "morgue," she heard typing coming from the business offices. However, when she looked to see who it was, she found the room empty.

Finally, she reports frequently hearing the sound of footsteps in the lobby near the "morgue."

C—'s Story

C— was an assistant publisher at the time of our investigation and told us she saw what she described as "floaties" while sitting at a desk in the choir loft in the early morning, around 6:00 am.

L—'s Story

Taking a momentary break from work, employee L— went to the back of the building and poured herself a cup of water and placed it on the counter. Ripples formed in the water as if something disturbed the cup. Thinking it was likely the result of footsteps causing vibrations to travel from the floor up the counter and into the cup, she attempted to recreate the effect by stomping around the area. It did not work, and the cause of the ripples remains a mystery. She reported this happening on at least three separate occasions.

K—'s Story

During renovations of the building, some of the contractors were tending to their construction work after hours. Nobody else was in the building. Nevertheless, they heard footsteps coming from the upper floor as well as the sound of doors closing. This report was relayed to us by a newspaper employee named K—.

S—'s Story

Graphic designer S— often found herself working late at the newspaper offices. While alone in the building, she experienced a feeling of being watched as well as cold drafts.

B—'s Story

B— was a manager at the newspaper. One day, he was alone in his office, tending to some paperwork, when he heard the sound of footsteps. Loud, heavy footsteps, as if someone were walking nearby wearing shoes with a heel. Surprised, he called out to the person. No response came. He went to investigate. No one was there. Indeed, he found the entire building empty. He was completely alone.

Perhaps a bit shaken by the experience, he nevertheless returned to his office to carry on with his business. But the paper he was working on was nowhere to be found. Somehow, even though he was alone in the building, his paper went missing. It was never recovered.

D—'s Story

While working at her job in classified sales, D— heard a strange voice. She described it as the voice of "the angry man who goes between the basement and the parsonage." According to a story she heard, this is the spirit of a man the police followed into the building and out to the parsonage in the 1980s or 1990s. The police ended up killing this man, which would perhaps explain the alleged spirit's angry demeanor.

Unrelated to the "angry man" story, D— also reports the feeling of a "distracted" presence in the building's lobby.

A—'s Story

Server rooms are like high-tech crypts. Staffed by IT technicians like A—, they're often isolated from the hustle and bustle of a company's day-to-day operations. Almost like churches or hospitals, they're maintained by a class of people with special training in the artifacts they house. Like many other rooms containing arcane equipment (albeit in this case high-tech rather than religious or mystical), they seem to lend themselves to a sort of creepy feeling.

A— was able to experience this feeling firsthand. While tending to his IT-related tasks in the server room, he felt a chill. This, he said, is a common occurrence and he paid it no mind, choosing instead to continue with his work. When he looked up, though, he saw the reflection of a face staring back at him from the screen of a computer monitor. Startled, he spun around to confront the individual, but the room was as empty as a pauper's purse.

Following this experience, A— attempted to communicate with the spirit and see what reaction he might receive. Remarkably, he said it worked and he was able to communicate with the ghost or entity. However, he refused to tell us what the spirit said. Which, to our minds, seems like the perfect set-up for a great horror story, but is more than a little frustrating from an investigative point of view.

Other Stories

Other stories (not told to us by any staff member during our interviews) contribute to the building's lore. Included among these are whispers that people can be heard moving about inside the building even when it's closed, that shadows are sometimes seen moving around the front desk, office equipment and papers move about on their own, and that loud parties have been heard in the basement. Most eerie of all is the claim that a ghostly choir can be heard singing inside the building.

The Inciting Incident

We were contacted by the staff of the Colorado Springs Independent following an incident that took place in 2006 and which was observed by multiple witnesses. Specifically, several staff members report that, in the first floor women's restroom, the hot water had been turned on when nobody was present in the room. Multiple witnesses report entering the room to find the water turned on all the way. They wanted us to see what we could learn about these strange occurrences.

As a brief aside, when word got out that we were working with the Indy, a local resident took offense and seemed to believe that our activities would invite some kind of demonic force into their community. Despite assurances that we are in no way involved with diabolic forces and certainly not aligned with the demons, this individual took to harassing some of the newspaper staff members to the extent that law enforcement had to get involved. By the time we arrived to investigate, he was sleeping off his delusions in the local jail.

Our Investigation

The Rocky Mountain Paranormal team arrived for our investigation on Saturday November 18, 2006. As we mentioned in one of our prefatory essays, we don't spend a lot of time on an investigation running around in the dark and scaring ourselves. Rather, we rely on silence, so as not to contaminate our own data, and a variety of monitoring devices to record the state of the investigation site.

In the case of The Indy, we set up video monitoring of the following locations:
- Main lobby (two cameras)
- Upper hallway (one camera)
- Lower ramp (one camera)
- Hallway leading to break room (one camera)
- Break room (two cameras).

Additionally, we placed audio recorders in the following locations:
- Main lobby (two microphones)
- Kitchen (one microphone)
- Computer/server room by ramp (two microphones).

Finally, we monitored the temperature in three locations:
- Main entrance
- Mid-building (in cubicles)
- Hallway in rear.

Figure 1.3. Our base of operations. Photo: Bryan Bonner.

The results from the temperature readings weren't terribly interesting. During the course of the investigation, the temperature rose from 67.6 degrees Fahrenheit to 74.6 degrees Fahrenheit in all of the monitored locations. Though we typically expect temperatures to drop as evening descends into night, the increase is easily explainable as resulting from our team's own equipment and body heat radiating into the building.

Throughout the evening, we took EMF (electromagnetic field) readings in several locations. In the main lobby, readings ranged from 0.0 to 4.0 milligauss. The higher readings were obtained near the front desk by the computer and the south side of the main entrance near the electrical boxes. In the upper hallway, we took readings in three locations down the hallway, obtaining readings between 1.75 milligauss and 3.0 milligauss. In the first floor break room, EMF readings in four spots were obtained, ranging from 2.0 to 12.0 milligauss. One location near the vending machine produced unusable data because the reading was beyond our meter's range. Finally, we took readings in one location in the hallway to the break room, obtaining EMF measurements between 12.0 and 14.0 milligauss.

Here we need to pause our discussion and explain a bit about electromagnetic fields. This isn't going to be a complete physics lesson, but something we've noticed watching a variety of other ghost hunting teams is that they will cite numbers from their EMF readings without providing any units. Electromagnetic fields are typically reported in units of either milligauss (1/1000 of a Gauss) or microTeslas (one millionth of a Tesla). However, these units, though both measure the strength of EMF, are not equivalent. Indeed, 1 milligauss is equal to 0.1 microTesla. Therefore, it's of vital importance to know which units are being reported. If someone just says a number, our interpretation could be off by a factor of ten or more if we don't know which unit they mean.

Furthermore, most people have no idea whatsoever what a normal EMF reading would look like and what an anomalous EMF reading would look like. The reality is that it depends on many variables and it's necessary to be specifically trained in EMF detection (not necessarily for paranormal purposes) before presuming to use these devices for paranormal investigation.

However, to provide a bit of context for this discussion (and for future discussions in this book), the background level of electromagnetic radiation to which we are all constantly exposed (both by our various devices and by the Earth's own natural electromagnetic field) tends to range between 0.1 and 4.0 milligauss. When electric devices are present, we expect to find elevated readings, but the reading is a function both of the absolute strength of the field and the distance between the meter and the field's source. An electric can opener at a distance of a few inches might produce an EMF reading of as much as 600 milligauss, but at a distance of a few feet, this number falls to only 2 or 3 milligauss. Similarly, a microwave oven at 30 cm (just under 1 foot) might produce an EMF in excess of 50 milligauss, but the same microwave at 1 m (a little more than 3.3 feet) might produce an EMF of only 4 milligauss. Hopefully those examples provide you with a frame of reference to understand our readings.

In the case of the Colorado Springs Independent, the EMF measurements we obtained were all within normal expected ranges and we did not measure any unusual variations between different measurements taken at the same locations.

After setting everything up and taking a first round of EMF measurements, our "quiet time" began at 12:30 am—just after midnight. This is an important part of our protocol and we only break it when we have very good reason to do so. In order to collect quality data, it's essential that everyone on the team knows where everyone else is, and that everyone is silent so as not to potentially create false readings on any of the recording equipment. Of course, staying silent and sitting in place all night wouldn't be tolerable, so there are regularly scheduled breaks in which people can move around, discuss anything they've observed, and take new readings.

The first part of the evening proceeded just as expected. As is typical of many of our investigations, we spent the evening sitting around, watching and listening as a whole lot of nothing was happening.

Just as it started to seem like another long night of nothing, our team members started to hear the sounds of music and singing. As you can easily imagine, hearing the sound of singing in the wee hours of the morning in a supposedly haunted former church is both exciting and likely to produce a Grade-A case of the heebie jeebies.

Both because the experience was anomalous and because the sounds of phantom singing were a part of the location's established ghost stories, we were fairly excited to have something we could investigate. After enjoying the creepiness of the ghost music for a moment, our team set about trying to determine possible sources of the sound.

Eventually, we found the culprit, and in this case, it's a fairly unusual one. As it turns out, the building is located near a busy freeway (I-25). Similarly, the main power plant servicing Colorado Springs, the Martin Drake Power Station (which burned coal at the time, then burned natural gas briefly before ultimately closing in 2022), was less than a mile from The Indy and between the newspaper building and I-25. Both of these—the highway and the power plant—are fairly noisy structures in their own right. But what we learned was even more interesting. When there is freeway traffic *and* the power plant is running at the same time, the combined sound waves enter The Indy's building and, due to the building's church architecture (designed specifically with choral acoustics in mind), end up producing a noise that sounds eerily like music and singing.

Never let it be said that the natural explanations of alleged paranormal phenomena are boring. No, there might not be a ghostly choir in the Colorado Springs Independent (at least not that we were ever able to observe), but the explanation, at least to us, seems equally intriguing. At around 5:45 in the morning, near the conclusion of our investigation, one of our team members again heard the singing while working in the break room. Even knowing the explanation, it was just as (or at least almost as) fun and spooky the second time around.

Unfortunately, we were never able to observe the hot water turning on in the women's restroom, so the thing that prompted our visit remains unexplained. Of course we do have some ideas. One possibility, naturally, is that some member of the staff, knowing about the building's haunted reputation, was playing a prank on his or her colleagues by turning the water on while no one else was in the restroom and then sneaking out unseen. It's entirely possible. It's also possible there could be a ghost causing the disturbance. Because we were unable to observe the phenomenon, we can only speculate.

Though we were unable to observe some of the claimed phenomena and were able to naturalistically explain another of them, that doesn't mean we didn't experience some unusual events for ourselves. One in particular is noteworthy.

At around 3:00 am, team members A and B[2] were taking a break, appropriately enough, in the break room. They were positioned such that B could see all the way down the hallway leading from the breakroom toward the rest of the building. Member A was turned toward B, facing in the opposite direction. During their conversation, B became rather excited and claimed he'd just watched a woman walk across the hallway and enter the women's restroom. He described a woman with shoulder-length brown hair and wearing a peach skirt. The team immediately got up and knocked on the women's room door. No one responded. After waiting for a moment, our investigators entered the room and found it empty. The women's room has only one entrance/exit, which was being watched, so no one could have snuck out a back door.

We played back the video, which included footage of the breakroom and the hallway. The only thing of interest we found on the footage was a rather strange expression on B's face when he saw the alleged woman. No woman was on the footage.

That experience remains unexplained. Was it a ghost? A hallucination? The result of some trick of light and shadow? There are many possibilities, and while we think some of them seem more likely than others, we're still left scratching our heads. Hopefully some day we'll be able to return to this amazing historic building and continue our investigation.

---

2   With rare exceptions, we've chosen not to identify which of our members performed which actions in these case files. Our intent is to keep the focus on the history, science, and ghost stories rather than on ourselves.

References and Further Reading

Archuleta, A. (n.d.). Colorado Springs Independent. *Colorado Springs Pioneers Museum.* <https://www.cspm.org/cos-150-story/colorado-springs-independent/#:~:text=The%20Colorado%20Springs%20Independent%20was,and%20thought%2Dprovoking%20articles%E2%80%9D.>

Central United Methodist Church (n.d.). Our History. *Central United Methodist Church.* <http://cumccs.org/OurChurchHistory>

Waters, S. (2012). *Ghost of Colorado Springs and Pikes Peak.* Charleston, SC: Haunted America, The History Press.[3]

---

3   The author of this book cites our investigation: "The Rocky Mountain Paranormal Research Society did an investigation of the building in 2006 and confirmed that it is haunted." This is, of course, an exaggeration of our findings. We had some explained as well as some unexplained experiences but would prefer to use a word like "interesting" as opposed to "confirmed."

2
The Not-So-Haunted Restaurant: Josephina's Restaurant

Josephina's Restaurant was an Italian eatery at 1433 Larimer Street in Larimer Square, part of Denver's LoDo district. Though now closed—the building is now home to cocktail bar Corridor 44 and award-winning Mediterranean restaurant Rioja—it was situated in the middle of a historic district and had a reputation for being haunted. Even today, when Josephina's happens to come up in conversation or at one of our public lectures, people who've been in the Denver area for more than a few years remember the Josephina's ghost stories.

This is also one of those cases in which our investigation itself ended up being more interesting than the ghost stories we were trying to investigate.

The History

The history of Josephina's, largely, is the history of Larimer Square. Though the Square is now home to commercial ventures and fine dining restaurants, the buildings predate its existence as a food district. Larimer Square is Denver's first locally protected historic district, receiving that designation in 1971 following Dana Crawford's campaign to save the block. That's unusual in that an entire district was marked for historic preservation rather than the individual buildings. But it makes a certain degree of sense, considering the long history of the block.

Indeed, Larimer Square predates even the existence of the State of Colorado. Colorado wasn't granted statehood until August 1, 1876 after its first bid for statehood was vetoed by President Andrew Johnson. Frustrating though that must have been, it seemed to work out well in the end, as the thirty-eighth state was grated admission to the union in the year the United States celebrated its 100th birthday and so forever became known as the Centennial State. Larimer Square, however, already had a history by that point. The Denver City Town Company was established in 1858. The very first cabin was constructed in that year to house General William Larimer. The location of that cabin was right in Larimer Square, where the Granite Building is currently located (and has been since 1882).

The Colorado Territory was established in the year 1861 (the same year as the beginning of the American Civil War). And as we've already mentioned, statehood was approved in 1876. Throughout this period, Larimer Square remained one of the major hubs of activity in the region. However, by the early 1900s, businesses began leaving the block in favor of the nearby 16th and 17th Streets. By the 1930s, Larimer Square had fallen into disrepair and was considered Denver's skid row. Things continued to deteriorate into the 1950s and 1960s, when the Denver Urban Renewal Authority targeted Larimer Square for…well, for "urban renewal," which likely would have cleaned up the area, but at the cost of losing the historic buildings in the neighborhood.

Enter Dana Crawford. Hers is a name you might not know off hand, but if you have any interest in Denver culture and history, you've undoubtedly benefitted from her work. Born Dana Hudkins in 1931, she moved to Denver in 1954 to work at a public relations firm after completing her education at Radcliffe College. In 1955, she married geologist John W. Crawford III and had four sons (some of whom continue to work with her on her historic preservation projects to this day). In the early 1960s, she took issue with the Urban Renewal Authority's plans for Larimer Square and sprung into action. She assembled a dedicated team of investors committed to the project of purchasing the Larimer Square buildings, one by one, and restoring them for new use. In 1971, her campaign succeeded in getting the entirety of Larimer Square approved for historic preservation and protection.[1]

However, Crawford's direct involvement with Larimer Square came to an end in 1986 when she sold the properties to the Hahn Company. In 1993, Josephina's owner Jeff Hermanson purchased Larimer Square (including, of course, the building in which he'd been operating his restaurant) and began operating it as Larimer Associates. Throughout the 1990s, some new buildings were added to the Square where parking lots had formerly existed, but the historic buildings remain largely as they have been for much of our state's history. Hermanson himself owned Josephina's Restaurant for about thirty years before it finally closed its doors.

In 2018, Hermanson and his partners sought amendments to the historic protections of Larimer Square. They wanted an exception to height limits in order to build two tall buildings, and to partially demolish some of the existing buildings to make room for the project. In response, the National Trust for Historic Preservation added Larimer Square to its list of 11 Most Endangered Places—the first time any Denver property has been listed in that document. Unfortunately for Hermanson and company but fortunately for historic preservationists, the plan did not go through, and in late 2020, the entire Larimer Square Historic District and its 22 buildings (along with a couple of other properties) were sold to Asana Partners, a North Carolina-based real estate investment firm.

The building that was to become Josephina's (and now split into Rioja and Corridor 44) has its own interesting history. Prior to the present building's construction, the property was home to the Clifford and Spencer corset shop. Known as the Frontenac Building, it was designed by Denver architect William Cowe in 1910, but its completion date is uncertain. Many of its occupants during its early years are unknown and it's difficult to separate fact from legend. It is known that at one point in its early years it was home to the Denver Leather Company on the first floor and the Frontenac Hotel from which it takes its name on the upper floors. According to local legend, while the Frontenac Hotel operated upstairs, the lower level became a speakeasy during prohibition. We have been unable to verify this piece of the building's history and, as we'll see, have plenty of reason for skepticism.

Now that half of Josephina's has become the Corridor 44 cocktail bar, ghost stories continue in its new incarnation. It is to the ghost stories that we now turn our attention.

Paranormal Claims

The main story you'll hear people tell of Josephina's (or now of Corridor 44) is

1   Colorado Women's Hall of Fame (n.d.). Dana Crawford. *Colorado Women's Hall of Fame.* <https://www.cogreatwomen.org/project/dana-crawford/>

History Colorado (n.d.). Dana Crawford. *Colorado Encyclopedia.* <https://coloradoencyclopedia.org/article/dana-crawford>

related to the building's reputation for having been a speakeasy during Prohibition. The owner of said speakeasy had a wife known as Amelia and it is her spirit who is supposed to haunt the location. The unnamed speakeasy owner and Amelia had a daughter named Ginger.

When Ginger reached the appropriate age for such things, she began dating a boy of whom her father disapproved. That seems like a fairly normal disagreement between fathers and daughters and had it ended there, there wouldn't be much of a story. But in this case, our speakeasy owner so strongly disapproved of his daughter's new beau that he hired a hitman to kill the boy. Unfortunately, the hitman was a little *too* successful in his job and managed to kill not only the boy but Ginger as well.

Amelia received news of her daughter's death while she was at the back of the speakeasy, in what later became Josephina's women's restroom. Her spirit is said to have remained in that place. This story comes courtesy of a psychic consulted by the restaurant's owners in the 1980s following their issues with apparently paranormal phenomena.

While the building was still Josephina's , the ghost was reported to break the restroom mirror and throw bottles at the bar. Now that the building is Corridor 44, the website www.hauntedplaces.org still maintains a listing for the ghostly activity, repeating Amelia's tragic tale.[2]

Our Investigation

When we reached out to the restaurant's management to inquire about a potential paranormal investigation at the restaurant, they invited us to dinner at Josephina's. Over that meal, our host regaled us with the location's ghost stories (which we've retold above). It's a great story, though admittedly some parts of it sounded a bit familiar to us.

Then came the confession.

The entire story, which, incidentally, was printed on a large poster in the building's entrance for all customers to read, was a complete fabrication. The management had made it all up as a way to help promote the business. And it worked. Recall our earlier essay regarding the profitability of owning a haunted business. Josephina's became a hotspot not only for those seeking a good meal but for those looking for haunted curiosities. And who could blame them? After all, it was the very same curiosity that landed us in the restaurant, too!

We were never provided with the source material that inspired the ghost story—if there was any—but it's worth noting that the overall structure of the tale is not entirely unfamiliar within ghostly lore. The story of a father killing the daughter's boyfriend is a common one. Typically the story ends either with the father accidentally killing his

2   Haunted Places (n.d.). Corridor 44 – Josephina's Restaurant. *HauntedPlaces. org.* <https://www.hauntedplaces.org/item/corridor-44-josephinas-restaurant/>

own daughter, as happened in the Josephina's tale, or with the daughter committing suicide after she discovers what happened. One of our favorite renditions of this kind of ghost story is the White Lady of Kinsale, a famous story from Irish lore, for example.

Obviously there is no direct line connecting The White Lady of Kinsale to the made-up ghost of Amelia at Josephina's Restaurant, but given that similar stories are not uncommon in paranormal lore, it seems likely that when the restaurant's management decided to write a ghost story to advertise their business, they created a kind of amalgamation of similar tales gathered either from books of folklore or any of a number of horror stories that have followed similar plots.

After this confession, our host asked if we still had any interest in investigating at the restaurant. Admittedly, knowing the ghost story was a complete fabrication does rather take the wind out of the sails a bit. However, we agreed to investigate anyway for a few reasons. For one thing, we just enjoy poking around a variety of places after hours. One of the advantages of paranormal investigation is that it gets us in to places most people don't get to see (or under conditions in which most people don't get to see them). Second, any opportunity for investigation is also an opportunity to further hone our skills and potentially develop new methodologies. Finally, just because the established ghost story turned out not to be real, there's nothing to absolutely guarantee that the building might not be haunted anyway. On the off chance we might nevertheless find something unusual, we agreed to an investigation and arranged a time to conduct it.

We arrived at Josephina's at around 1:00 am and asked to see the manager to announce our presence and start setting up. The late hour was not an anomaly. Josephina's also operated as a bar which would not close until 2:00, so we were, in fact, dutifully early for our appointment. The manager on duty came out and, of course, had no idea who we were or why we were there. In retrospect, that might have been an omen of things to come. But miscommunication is not entirely uncommon in this field, so we proceeded as normal. The manager called the owner who confirmed that we really were supposed to be there. I'm sure that was a relief to the manager who must for a few moments have been worried about the sudden appearance in the middle of the night of strange people hauling what appeared to be some unholy hybrid of a scientific laboratory and a television studio.

The bar was meant to close at 2:00 in the morning. However, there were some staff issues and the last customer wasn't out of the building until 2:30. Not a major inconvenience, but perhaps another sign as to the sort of evening/morning that was in store for us.

As soon as this last customer left, we began setting up our monitoring equipment and collecting baseline readings (EMF, temperature, and so forth) throughout the restaurant. After setting up, we went to let the manager know we were ready to begin our work. No one, however, was to be found. They'd left us completely alone in the building. This is not entirely unusual—to our eternal surprise, people seem quite willing to just let us run amok in their homes or places of work without any kind of security or supervision—but usually they at least announce their departure.

During one of our sweeps of the building, we noticed some unusual activity, but not, unfortunately, of a paranormal variety. Through the rear kitchen window, we were able to observe a parking lot that seemed rather active as the local bars were letting out and their patrons returning to their vehicles (hopefully sober enough to drive safely, though we have our doubts). The one incident of note during this exodus from the watering holes occurred when a very drunk man crawled into the back seat of a car. As it turned out, the car did not belong to him. The owner returned a few minutes later and received quite a fright upon discovering an unconscious drunk in the back seat.

Later in the evening, the parking situation again became the subject of our attention as two of our team members' cars were towed. The nearby street parking area was unclearly marked and this resulted in the cars being impounded.

Monitoring of the restaurant continued with no observed anomalies.

At about 5:45 am, we heard the front door of the building open, which was a bit startling. Two individuals entered. It turned out they were the morning cleaning crew, rather surprised to see the building occupied by a team of researchers who must by this hour have looked like the walking dead somehow got hired as a film crew. Of course, like the manager before them, they had no idea who we were, what we were doing, or that we were supposed to be there. Our attempts at explanation were not the most effective as neither of them spoke a word of English. However, our Charades skills were on point and we were eventually able to mime our way through enough of an explanation to get their boss on the phone who was able to sort the whole thing out.

At that point, we decided to call it a day, packed up our things, and went home.

We weren't able to collect any kind of data that could have suggested paranormal activity. And in light of the confession that the original ghost stories were a hoax, we were happy to chalk this one up as pretty close to definitively "not haunted." Though our experiences throughout the evening do leave some questions open as to whether the building may instead be cursed. Regardless, this went down in Rocky Mountain Paranormal history as one of the most surreal investigations for completely non-paranormal reasons. We may call it the "not-so-haunted restaurant," but we have a certain fondness for telling its story anyway.

References and Further Reading

Authentic Vacations (n.d.). Haunted Places: Who is the Ghostly Lady of Kinsale? *Authentic Vacations*. <https://www.authenticvacations.com/the-white-lady-of-kinsale>

Colorado Women's Hall of Fame (n.d.). Dana Crawford. *Colorado Women's Hall of Fame.* <https://www.cogreatwomen.org/project/dana-crawford/>

Crawford, D. (n.d.). Bio. *DanaCrawford.net*. <http://www.danacrawford.net/bio.html>

Goodstein, P. (1996). *The Ghosts of Denver: Capitol Hill*. Denver, CO: New Social Publications.

Haunted Places (n.d.). Corridor 44 – Josephina's Restaurant. *HauntedPlaces.org*. <https://www.hauntedplaces.org/item/corridor-44-josephinas-restaurant/>

History Colorado (n.d.). Dana Crawford. *Colorado Encyclopedia*. <https://coloradoencyclopedia.org/article/dana-crawford>

History Colorado (n.d.). Larimer Square. *Colorado Encyclopedia*. <*https://coloradoencyclopedia.org/article/larimer-square*>

Warner, J. (2007). Haunted Happy Hour. *Westword*. <https://www.westword.com/arts/haunted-happy-hour-5095547>

3
The Voice at Victor:
The Victor Hotel

The Victor Hotel is a four-story Victorian brick building in Victor, Colorado, a small mountain town in Teller County, southwest of Colorado Springs and near Cripple Creek. Visiting the hotel is a lot like stepping back in time. Though the hotel has been renovated and offers plenty of modern amenities, it has maintained its historic ambiance and is listed on the National Register of Historic Places.

Our investigation began when we were asked to be part of the KOA Radio Zoo Boo tour for a Halloween show featuring hosts Scott Hastings and Dave Logan. The show was to be broadcast live on location from the Victor Hotel for two days, giving us ample opportunity to explore this unique building.

Figure 3.1. The Victor Hotel's (allegedly) haunted elevator. Photo: Bryan Bonner.

The History

As is the case with many historic Colorado properties, the history of the Victor Hotel is intimately connected with the mining industry. On July 4, 1891, William Scott Stratton discovered the aptly named Independence Lode near what is now the town of Victor. Between 1893 and 1899, an estimated 200,000 Troy ounces (about 13,700 pounds or 6200 kilograms) of gold was removed from the mine. This made Stratton the first millionaire of the Cripple Creek Mining District by 1894. The same gold in 2022 would be worth more than $343 million. This is not a small amount of money we're talking about.

The town of Victor was founded in 1891, almost immediately after Stratton's discovery and named for the Victor Mine (itself thought to be named for an early settler named Victor Adams). In 1892, Warren Woods arrived in Victor with his sons Frank and Harry Woods. Warren oversaw most of the Woods' enterprises, delegating a variety of managerial tasks to his sons. Frank oversaw the Woods Investment Company and Harry was a newspaper man until the family's relocation to Colorado. Upon their arrival, the Woods family established the Mt. Rosa Mining, Milling and Land Company. Partly due to the Woods' development projects, the town was growing and officially became a city in 1894, the same year the Woods brothers discovered gold while digging the foundation for one of their buildings. This became the Gold Coin Mine.

While Cripple Creek is now a much more famous mining town than its smaller neighbor, Victor became known as the "City of Mines." Together, mines in Victor and Cripple Creek produced some 21 million Troy ounces of gold, the value of which in today's market would be over $36 billion.

The original Victor Hotel and Restaurant was the building project that resulted in the Gold Coin Mine discovery. Nevertheless, the Woods were able to construct their building in 1894. Their goal was to establish a hotel and restaurant to provide comfort for travelers on the Florence and Cripple Creek Railroad which was founded the year prior. As a side note, the Railroad went out of business in 1915, but Phantom Canyon Road, following much of its original route, is open to public traffic during the summer months.

The building was quite the impressive accomplishment. It was a two-story wooden frame building featuring a cone-shaped tower and enclosed batteries. Arguably the inside was even more impressive—this marvel of cutting edge technology and engineering even featured electricity, a rarity at that time.

Unfortunately, not all good things last. On August 21, 1899, a fire broke out in Victor, then home to around 18,000 residents. The fire is thought to have begun at the Merchants' Cafe shortly after noon but was fanned by winds from the south and within five hours had destroyed the entire business district, including the Victor Hotel and Restaurant. The next day's *Telluride Daily Journal* described the damage:

Town of Victor Wiped Out by Fire. About 8 o'clock last evening
L. C. Hunt, manager of the Telluride house of the Tompkins Hardware Co., received a telegram from the manager of the Victor house
of the same firm which read: "Store and warehouse burned down.
Entire business portion of Victor gone." Between the store and warehouse three blocks intervened, and to the telegraph office was four
blocks in another direction. The telegram was dated at the Goldfield
office, indicating that the Victor telegraph office had been destroyed.
This will give an idea of the scope of territory devastated by the fire.[1]

But the people of Victor, including the Woods brothers, were not so easily defeated. The town was quickly rebuilt, this time using brick and stone frameworks instead
of wood.[2] If you visit the town's historic buildings today, you'll find that many of them
were built not when the town was founded, but in 1899, following the fire.

The same is true of the Victor Hotel. The Woods Brothers rebuilt their hotel
across the street from the original location. However, even the new location had been
touched by the fire. The Woods owned a shaft house where the current Hotel is located
which was also consumed by the fire. The new Victor Hotel is a four-story brick and
stone building completed on December 24, 1899, and remains the largest building in
Victor. Upon opening, it was home not only to the Victor Hotel's lodging rooms but
also the Woods brothers' investment company, the First National Bank, and other re-
tail spaces. The first floor consisted of storefronts and the upper floors consisted of
lodging rooms and office spaces.

The hotel also featured a bird-cage elevator to carry passengers from floor to
floor, which proved useful during the winters. When the ground was frozen in the
winter and the local cemetery therefore unable to dig new graves, bodies were trans-
ported to the hotel's fourth floor, which had become a hospital in 1906, for storage
until the spring thaw. A practical solution to a problem, no doubt, but also excellent
fodder for plenty of ghost stories.

Unfortunately, the Woods brothers' wealth and fortune were not to last indefi-
nitely. In 1903, bank examiners found the bank insolvent and forced its closure, after
which the Woodses sold the building. It subsequently became the Citizen's Bank of
Victor and then, under the ownership of A. E. Carlton, the City Bank. However, the
Great Depression was hard on the building. The City Bank closed along with many

1   Quoted in: Western Mining History (n.d.). Victor, Colorado. *Western Mining
History.* <https://westernmininghistory.com/towns/colorado/victor/>

2   Having studied the history of numerous locations, particularly those in and
near Colorado, we've discovered that most early structures were wood and a substantial portion of them burned down and were eventually replaced with brick and stone.
We often remark that nearly the entire state of Colorado burned at some point in its
early history.

of the other businesses. In the 1930s the building was home to a photography studio and café. In the 1950s and early 1960s it was a gift shop, restaurant, and soda fountain. But all good things must come to an end, and it closed in the 1960s and fell into disrepair over the course of the following years.

The Victor Hotel's modern life began in 1991 when it was purchased by Victor Hotel, LLC and a restoration process began under the supervision of Marjoe D. Bandimere. This renovation was completed in 1992 and it is now open to the public, operating both as a hotel and event space as well as a destination for history buffs and paranormal thrill seekers.

The philosophy of its renovation was to include all the modern amenities while still maintaining its historic character. It now has all of the modern conveniences we expect from any hotel, but still features its bird-cage elevator, bank vault, and original (however restored) woodwork.

Paranormal Claims

Of course, given the building's history, there are sure to be plenty of ghost stories. Something we've observed over the years is that almost every old hotel and almost every old hospital has some spooky legends associated with it. In the case of the Victor Hotel, everything seems perfectly aligned to guarantee some good spooky lore. After all, it's an old hotel that once contained a hospital and sometimes acted as a makeshift morgue. What more could we possibly ask of a historic building?

Some visitors and ghost hunters claim to have seen apparitions that look like doctors and patients wandering around the fourth floor. By some reports, these spirits are sometimes missing limbs or even their heads. It is not clear whether these appendages are missing because spirits have no need of them or because the people are meant to have lost them during their lives.

Ghost stories have also been told about the basement, elevator, third floor, and kitchen. In the kitchen, staff have reported utensils being thrown with no apparent means of propulsion. Others have reported phenomena fairly typical of alleged hauntings including the sounds of footsteps, drafts of cold air, and the appearance of misty figures.

By far the most prominent ghost story, however, is the tale of Eddie. During the days of the gold rush, the Hotel had served as a boarding house for miners. One such miner was Eddie, whose surname is not included in any of the stories. According to the legend, he'd been staying in room 301. One day, he called the elevator. The gate opened and Eddie stepped through. Unfortunately, he didn't realize the car had not yet arrived and so fell down the empty elevator shaft to his death. His spirit is said to haunt either the elevator, room 301, or both.

Our Investigation

As a rule, we conduct investigations in silence. We like to be the only people present (or at least the only people other than the property's owners) to minimize the possibility of contaminating our own data. That's not to say we don't like doing public events. We love doing them. When not investigating, you'll usually find us telling ghost stories to audiences large and small. It's just that the public work and the investigative work have completely different demands. However, sometimes an opportunity presents itself and we have to bend our usual protocols. Our work at the Victor Hotel was just such an occasion.

We were approached by the KOA radio station. Around Halloween they were doing a programming event called the Zoo Boo tour. As part of this series, they planned to do a two-day live broadcast from the Victor Hotel. We were called in to perform an investigation (at least to the best of our ability under those circumstances) as well as to appear as guests on the show to discuss the hotel's ghost stories and the outcome of our investigation. Of course we agreed, because even for us, it's still not quite *every* day we get the opportunity to spend two days telling ghost stories on the radio live from a haunted hotel in a historic mining town.

Prior to conducting our research on site, we began with some historical review of the property. We have attempted to locate any public records corroborating the story of a miner named Eddie who might have stayed in Room 301 at the Victor Hotel and might have died on the premises. We found numerous sources that repeated the ghost story but have, at least so far, been unable to find any official records documenting the story.

Fortunately, the hotel was closed to the public at the time of our investigation, so we didn't need to worry about the whole thing turning into a complete circus. The presence of a radio crew conducting a live broadcast complicated matters for us, but it was still controlled enough that we were able to do our work. We arrived on the afternoon of Halloween.[3] Both our team and the radio show's team quickly toured the facility and then began setting up.

The radio hosts Scott Hastings and Dave Logan decided to stay in Room 301. If you're going to do a haunted radio show, you need to be in the haunted room. That makes sense. We placed some monitoring equipment in 301 and the adjoining rooms.

---

3   That sounds like the perfect time for some ghost stories. In reality, we rarely work on Halloween itself, preferring to take that day off so we can enjoy our holiday and scare the various trick or treaters in our respective neighborhoods. We're a lot like the demons and vampires in *Buffy the Vampire Slayer* in this regard.

Figure 3.2. A guest room loaded with monitoring equipment. Photo: Bryan Bonner.

As the evening began, we settled into our routine of monitoring the video and audio feeds from our recording equipment throughout the hotel. Unfortunately, the video feed captured the radio hosts beginning to disrobe in their room. More unfortunately, this was displayed on a monitor directly facing the entire remainder of the crew. We quickly turned the monitors off, just in time to avoid undue embarrassment. There's never a dull moment, apparently, in paranormal investigation. But this does further illustrate why we don't make our investigations open to the public—even when it's just ourselves and a professional radio crew, situations can quickly get out of our control.

Once they'd completed their evening broadcast, the radio crew retired to a local bar and restaurant outside of the hotel. This was welcome news for us because it allowed us free run of the hotel to conduct our investigation in the kind of isolation to which we're accustomed. During this time, we monitored several locations throughout the hotel for a period of some hours without anything unusual occurring.

Then one strange thing did happen. One of our team members was working in the stairwell between the third and fourth floors of the hotel when he heard a voice. Immediately, he reported the anomaly to the rest of the team and we began to investigate. At the time, our investigator was carrying a camcorder capable of recording audio and video as well as a mini-disk audio recorder. We played back these recordings and found that only the audio recorder captured the anomalous voice. The voice itself was a distinct male voice whispering: "Who are you? Who are you? Good night." Following the whispering voice, the audio also contains the sound of loud footsteps. This last part, however, was not at all paranormal. These footsteps were those of our investigator running back to rejoin the rest of the team elsewhere in the hotel.

One thing we'll note is that this is distinct from the vast majority of ghostly recordings we've encountered from other teams throughout the years. It's common practice to attempt to record EVP, or electronic voice phenomena (essentially ghost voices), on audio recorders. Typically, the results are garbled sounds that are only intelligible if one has been psychologically primed to hear a specific word or phrase. In this case, by contrast, the voice is clear, distinct, and articulate. You don't need a written caption to be able to make out what it says.

Since a book can't contain audio, you'll have to take our word for the tape's quality for the moment. If you want to hear the recording for yourself, we're more than happy to play it for anyone who asks for it at one of our public lectures or events.

When the radio crew returned to the hotel, we played the audio for them and gave them a copy to include in their broadcast the following morning, which they did. Repeatedly. At some point, the radio crew went to bed (again in room 301), and our team stayed up to continue trying to figure out the source of the voice.

At one point while listening to the audio again to try to find some clue as to its source, we suddenly heard a loud commotion coming from the direction of Room 301. We immediately ran to the room. When we arrived, the door swung open and we were greeted by the sight of one radio host on the floor, struggling to hastily dress himself. Meanwhile, the other tripped over his colleague while trying to simultaneously put on his shirt and run out of the room. Both of the men kept exclaiming "It's in here with us."

Unfortunately (or perhaps quite fortunately indeed for the two men involved), there was no otherworldly entity in their room. They'd simply heard us (in the next room over) playing back the recording from earlier and thought the ghost had come to their room.

We never were able to determine the source of the mysterious voice. One thought that occurred to us, in light of the fact that our friends in the neighboring room were so readily able to hear our replay of the recording, was that sound may travel easily through the hotel, and we may have simply recorded someone speaking in another part of the hotel. However, this is not an entirely satisfactory explanation because as far as we knew, we were supposed to be alone in the hotel, and we know that none of our team members uttered those words.

Anomalous sounds can often be caused by other occurrences in the surroundings—consider the mysterious singing at the Colorado Springs Independent discussed in Chapter 1—but the clarity of the whispering voice in this recording leads us away from this kind of explanation. It's possible for random noises, under the right conditions, to sound somewhat like a human voice, but we'd consider it quite extraordinary for such random noises to produce such a clear and articulate voice as the one we recorded.

Does that mean we recorded a ghost? Not necessarily. Since we can't explain it otherwise, we have to remain open to that possibility, but a single unexplainable (or perhaps we should say "unexplained"—there's nothing to suggest that the recording is unexplainable in principle; we just haven't figured it out *yet*) recording does not constitute proof of the supernatural.

Whatever the source of the mysterious voice, our experience at the Victor Hotel was certainly a memorable one, and the hotel itself is an excellent piece of history worthy of our attention.

References and Further Reading

Colorado Urban Legends (n.d.). The Haunted Victor Hotel. *Colorado Urban Legends.* <http://www.coloradourbanlegends.com/haunted-victor-hotel/>

Haunted Colorado (n.d.). Victor. *Haunted Colorado.* <http://www.hauntedcolorado.net/Victor.html>

Victor Colorado (n.d.) Victor History. *Victor, Colorado: City of Gold Mines.* <https://victorcolorado.com/history.htm>

Victor Hotel (n.d.). 19th Century Charm, 21st Century Amenities. *Victor Hotel.* <https://victorhotelcolorado.com/about-victor-hotel.cfm>

Western Mining History (n.d.). Victor, Colorado. *Western Mining History.* <https://westernmininghistory.com/towns/colorado/victor/>

4
America's Oldest Press Club: The Denver Press Club

As the title of this chapter says, the Denver Press Club is America's oldest press club, tracing its history all the way back to 1867. However, the Press Club building located at 1330 Glenarm Place in Denver, CO dates back to 1925. In fact, that whole neighborhood seems to have something of a haunted reputation—stay tuned for a future volume of this series in which we discuss the Denver Firefighters' Museum just across the street from the Press Club. And it's still operating today, offering membership not only to professional journalists but also to members of the public for use as an event center and social gathering place.

Of course, it's not just members of the fourth estate who inhabit the building. It's home to more than a few ghost stories, and many people who've spent their share of time in the building report a variety of strange occurrences.

Figure 4.1. The Denver Press Club. Photo: Bryan Bonner.

The History

The first hint of what would eventually become the Denver Press Club began in 1867, when journalists started holding regular social gatherings. These pioneers of the fourth estate put aside their political and professional rivalries (often quite bitter) in the spirit of camaraderie, sometimes for a drink or poker game and sometimes just to talk. Lacking a space of their own, they held these early meetings in the basement of Wolfe Londoner's grocery store on Larimer Street.

The Denver Press Club was officially incorporated as a legal entity a year later in 1877. However, the popularity of the club soon outgrew the grocery store basement and the quest for a new location began. In the meantime, meetings were held at a variety of locations, primarily hotels, including the Brown Palace (which has its own ghost stories about which you'll be able to read in a future volume of this series).

In March of 1905, Edward Keating, then President of the Press Club, organized a meeting of some fifty newspapermen at the Albany Hotel to discuss the Club's reorganization. They had an ambitious vision. Instead of simply being a social club for journalists, the Denver Press Club should be an unprecedented source of entertainment for journalists but also for statesmen, captains of industry, and so forth. Honorary memberships were even granted to two United States Presidents: Theodore Roosevelt and William Taft. Of course other Presidents have spent time in the Press Club, including Warren Harding and Woodrow Wilson, but only Presidents Roosevelt and Taft were honored with the gold and silver membership card.

In 1906, the International League of Press Clubs came to Denver and elected Denver Press Club President Edward Keating as their president for the year.

The close relationship between the Press Club and the political movers and shakers is perhaps best illustrated by the fact that in 1908, the Press Club offered its location as headquarters and its staff as organizers for the National Democratic Convention. Unsuccessful Presidential candidate William Jennings Bryan (he lost in 1908 to Roosevelt's preferred successor, William Taft) was a frequent visitor to the Press Club.

By 1925, the Press Club could no longer continue its existence hopping from one hotel to another and finally obtained its permanent clubhouse at 1330 Glenarm Place. The building, where the Press Club remains to this day, was designed and built by architects Merill H. and Burnham Hoyt (who also designed the Denver Public Library and Red Rocks Amphitheater, the famous live music venue, among dozens of other Denver area landmarks). The cost of the building was $50,000 (equivalent to about $850,000 in 2022). Funding was provided primarily through Francis Kirchof's sales of *Who's Who in the Rockies* booklets which allowed local personalities to have their biographies printed in the publication for a fee.

Amazingly, the building has remained in the Press Club's hands ever since, though it has also received protection as a historic landmark. In 1986, the Denver Landmark

Preservation Commission honored the building with protection as a Historic Land-mark. A similar honor, recognizing the building as a "significant historical place in journalism," was granted by the Society for Professional Journalists in 2008. And as re-cently as 2017 the building was recognized by the National Register of Historic Places.

Over the years, plenty of famous personalities have joined the Press Club. Among them are Carl Akers, Bob Palmer, Eugene Field, Fredrick G. Bonfils, and Damon Runyon. If we count visitors rather than only members of the Press Club, we have to include a long list of Presidents, statesmen, public figures, and even the members of the Rocky Mountain Paranormal Research Society who are, of course, the most im-portant and distinguished of all the Press Club's many visitors over the years.

One thing that's remarkable about the Denver Press Club's history is that, though it has touched all of history since its founding, it has remarkably little history of its own. Likely this is because the building has remained in the same hands throughout its entire history, so the various comings and goings that haunt the histories of so ma-ny other old buildings are all but absent.

Today, the Press Club features a bar and restaurant on its main floor, a meeting and events hall on the second floor, and billiards and card rooms in the basement. The latter is named for a local artist, Herndon Davis, whose name frequently comes up in discussions of both local history and ghost stories. There is a mural on the game room wall painted in 1945 by Davis, who is also credited with the famous "Face on the Bar Room Floor" painting at the Teller House in Central City. While the latter is a subject for a different conversation, the Press Club mural is a wonderful piece depicting the best-known journalists of the day hard at work. Right below that mural is a card table, and that leads us directly to the ghost stories.

Paranormal Claims

Rocky Mountain Paranormal was not the first paranormal investigation group to visit the Denver Press Club. One of the challenges we constantly face in our line of work is to separate what we might call the "naturally occurring" lore surrounding a location (which comes from the venue's staff or customers and grew up organical-ly) from that which was added on post hoc by psychics or paranormal investigators who've gone through the location. Fortunately in the case of the Denver Press Club, unlike so many others, the self-professed psychics who preceded us to the location seem largely to have repeated pre-existing stories rather than substantially adding to or altering them.

While the Press Club has more than its share of alleged hauntings, one of them is more prominent than the rest, and dovetails perfectly with the history of Denver journalism. Working in journalism isn't always an easy job. While we don't think it's con-troversial to suggest that the fourth estate today isn't what it was even a few decades ago, the fact of the matter is, these people work incredibly hard to bring the news to the masses. And people who work hard also tend to play hard.

One of the favorite pastimes of the newspapermen of old was the game of poker. That's probably still true today, but back in the day it was arguably even more serious. Legends and lore of the Wild West are filled with stories of fortunes won and lost on the turn of a single card, not to mention all the tales of poker games gone wrong that may have ended in bloodshed. The Denver Press Club doesn't have any of those bloody stories, but they do have one legendary poker game.

According to the legend, there was a poker game in 1925, when the Press Club was set to relocate to their new permanent clubhouse on Glenarm. Four club members in particular were so immersed in their poker game that they refused to stop playing cards when it came time to move from the hotel to the new clubhouse. Indeed, the story says that they had to be moved with the furniture, continuing their game in the moving truck. Historic legend meets paranormal legend here because it's said that their ghosts continue the Phantom Card Game in the basement game room of the Denver Press Club.

Figure 4.2. Do ghosts carry on an eternal card game in this room? Photo: Bryan Bonner.

A self-professed psychic and her assistant were recruited by the Press Club to check out the location. The psychic claimed to have sensed not only the card players but also a woman who'd been murdered some 25 years prior, a hanged person, and a reporter standing lookout at the club. The psychic and her assistant proceeded to perform a "cleansing" of the location. She reported that the vast majority of paranormal entities were thus removed but the old newspaper men still refused to abandon their card came and remained behind.[1]

In addition to this group of four ghostly card players in the basement card room, the site is host to a number of other reported paranormal activities or occurrences.

The upper floor office of the Press Club is said to be haunted by the spirit of a woman who was murdered and then set on fire in the dumpster near the office window behind the building.

Some people report seeing a spectral woman descending the main staircase to the bar.

Numerous people report objects (glasses, bottles, chairs) throughout the building, and particularly in the bar, moving of their own accord.

Many people report hearing mysterious voices or sounds of footsteps throughout the building, even when they believe themselves to be alone.

Visitors and staff report unexplained temperature changes as well as "bad feelings" in the upstairs office.

Human faces have been seen in the mirror at the bar in front of the building.

Figure 4.3. Is this bar haunted by newsmen of the past? Photo: Bryan Bonner.

In July of 2004, a board member of the Denver Press Club was engaged in conversation with the general manager of a large corporation when both individuals witnessed a rather remarkable occurrence. A wine glass floated four feet above the bar and then fell straight down to the floor. Upon impact, rather than shattering, it merely broke in half.

In September of 1997, the Center for Scientific Investigation into Claims of the Paranormal (CSICOP; now the Center for Inquiry or CFI), a skeptical organization, published an article challenging the paranormal stories: "With the spirits still floating around, and the Press club 'Documenting' the ghost with fuzzy infrared photographs, perhaps one of the club's members will recognize the importance of this discovery

---

1   Fitzgerald, M. (2011). Ghost in Denver Press Club? *Editor & Publisher.* <https://www.editorandpublisher.com/stories/ghosts-in-denver-press-club-p8,141523>

(The Psychic Investigation), call in a team of real scientists with infrared detectors, find proof of the hereafter, and get the Pulitzer for investigative reporting after all.'[2]

However, prior to our work at the location, and despite being visited by numerous psychics and ghost hunters, there had never been a proper scientific inquiry, by either the believers or the skeptics, into claimed paranormal activity at the Denver Press Club.

Our Investigation

We began our investigation, as we usually do, by attempting to verify as much of the history of the location as possible. Of particular interest to us were the stories of the card game moving along with the furniture and the woman who is said to have been killed and lit aflame behind the building.

Unfortunately, records are sparse. Though we found plenty of people telling the story of the card game moving along with the building, it's not the sort of thing that one finds reported in public records, so all we really have is hearsay. It may or may not be true, but it certainly makes for a good story. It's just plausible enough to be believable, yet just extraordinary enough to be memorable. Of course that makes it the perfect kind of story to be consistently retold regardless of its veracity. Ultimately, unless someone can produce a photograph of the event (which is unlikely—photography certainly existed at the time, but this was before everyone carried cameras in their pockets and took snapshots of everything that happened in their lives), the best we're likely to be able to do is classify it as unverifiable hearsay.

The case of the murder is more likely to be the subject of public records. Despite our best efforts, however, we have been unable to find any records of a murder matching the description of the one from the ghost stories. That doesn't mean we've definitively disproved the story, but it does make it seem far less likely to be true. Particularly that grisly a murder seems like it would have been the subject of a newspaper article and should have been recorded in police records. We have been unable to locate any such articles or records. Given that it supposedly took place in the journalists' own clubhouse, we're pretty sure at least one of them would have thought to write about it.

We were, however, able to discover some of the history of the property prior to the construction of the Press Club. The Glenarm Street property prior to the Press Club being built was the location of a boarding house. This building was in complete disrepair, and ripe for replacement. The boarding house residents lived without the benefit of indoor plumbing *or* an outhouse (which beggars the imagination). Further, some of the rooms were used to house animals the residents used both as pets and as livestock. It was not a pretty place to live, despite its outward appearance as a delightfully creepy big old house.

2   Emery, C. E, Jr. (1997). Dark Skies Uses Pseudo-Sagan To Recast Astronomer's Motives. *Skeptical Inquirer, 21*(5): 21.

Interestingly, none of the Press Club's ghost stories, at least as far as we've been able to discover, extend as far back as the boarding house. It seems like the sort of property that would be ripe for paranormal rumor, but the ghost stories seem to begin only with the new building once the property was acquired by the Press Club.

Our first on-site investigation, though it managed to get some interest from the press (whether due to the inherent interestingness of our work or simply because it was at the Press Club is a matter for someone else to decide), wasn't terribly interesting. None of the specific phenomena that have been reported over the years manifested in our presence. Certainly nothing as spectacular as the physical manifestation of a Phantom Card Game or a floating champagne glass.

One area of some interest was the boiler room. When we entered the room, we found it to be home to an uneasy sort of feeling. Of course "feelings" aren't the kind of scientific data we're looking for, but one thing we've learned is that when that sort of feeling happens, it's often worth paying attention to because there could be something more concrete we can actually measure.

In this case, we were correct to investigate further. We found that the boiler's pilot light had gone out. Therefore, there was a slow leak of gas into the room. In addition to the severe fire hazard posed by a gas leak, exposure to high levels of natural gas can produce a number of physical symptoms which may be confused for paranormal phenomena, including unexplained headaches, nausea, difficulty concentrating, dizziness, loss of memory, and more. Indeed, we've added air quality monitors and combustible gas "sniffers" to our tool kit to account for precisely these kinds of phenomena during our investigations.

Fortunately in this case, no harm was done. The gas was turned off pending repair and no other unusual phenomena (whether paranormal or mechanical) were observed. Though in the aftermath we were called "ghost busters turned gas busters," so at least we got a new title out of the arrangement. And in our small way, we take a bit of credit for saving the Press Club, albeit from a fire hazard rather than a ghost. It may not be what we went there looking for, but we'll take it.

References and Further Reading

Emery, C. E, Jr. (1997). Dark Skies Uses Pseudo-Sagan To Recast Astronomer's Motives. *Skeptical Inquirer, 21*(5): 21.

Fitzgerald, M. (2011). Ghost in Denver Press Club? *Editor & Publisher.* <https://www.editorandpublisher.com/stories/ghosts-in-denver-press-club-p8,141523>

History Colorado (n.d.). Denver Press Club. *History Colorado.* <https://www.history-colorado.org/location/denver-press-club>

Kania, A. J. (n.d.). History. *Denver Press Club.* <https://denverpressclub.org/about/denver-press-club-history/>

5
The Work of Ghouls: Cheesman Park

Denver's Cheesman Park consists of 80 acres of parkland near the Capitol Hill neighborhood. Though the larger Cheesman Park neighborhood, including a substantial residential area, extends further, the park proper is approximately bounded on the north and south by 13th and 8th Avenues, respectively, on the west by Humboldt Street, and on the East by the Denver Botanic Gardens. One of Denver's worst kept secrets (in no small part due to our own lectures consistently reminding people) is that before Cheesman Park was Cheesman Park, the land was the Mount Prospect (also sometimes called Prospect Hill) Cemetery, Denver's first cemetery. Adding to the creepy history surrounding the park, not all of the graves were moved when it became a park. The fact that people enjoy their recreation just a few feet above ground in which bodies are still buried lends itself to plenty of ghost stories, and to one of Rocky Mountain Paranormal's most triumphant activities.

Figure 5.1. Cheesman Park. Photo: Bryan Bonner.

The History

Though historical research is part and parcel of all of our activities and we maintain that reminding people of history is both necessary for and one of the primary social functions of paranormal investigation, Cheesman Park is perhaps unique in the degree to which a full understanding of the Park's history is necessary for an understanding of its spooky side. In the case of, say, a haunted mansion (any of the ones from our case files will do), it's easy enough to understand the ghost stories merely on their own terms. The history adds another layer of understanding, but one can grasp the ghost story without knowing the life stories of those involved. Cheesman Park is different. The ghost stories, as we'll see, are actually somewhat nebulous and often ill-defined, even by ghost story standards. The real story of Cheesman Park *is* the historic story.

The story of Cheesman Park begins long before it was a park. Indeed, it begins even before the land was the Prospect Hill/Mount Prospect Cemetery. In order to understand Cheesman Park, we need to go back substantially further in time. Long before Denver was a city, before Colorado was a state, and certainly before anyone ever thought of putting a park on the land in question, the land belonged to the Arapahoe, who used it as one of their own burial grounds.

The land on which Denver sits did not originally belong either to Denver or to Colorado (neither of which had yet been founded), nor even technically to the United States. It was part of the land reserved for the Arapahoe as part of the 1851 Treaty of Fort Laramie. That treaty, however, was doomed to failure, as it was almost immediately followed by continued American expansion into the West and the Pikes Peak Gold Rush.

In November of 1858, General William Larimer (for whom Larimer Square is named) established the Auraria mining colony[1] in what was then part of the Kansas Territory. The same year, the site's name was changed to Denver. Larimer was attempting to curry favor from Kansas Territorial Governor James W. Denver, though he was not aware that by the time he named the budding city in Denver's honor, Denver had already resigned as Governor and could no longer provide any political favors.

1   The name "Auraria" is taken from the Latin word for gold, "aurum," which is also why gold has the symbol "Au" on the Periodic Table of the Elements. The Auraria mining colony actually still exists as a neighborhood in Denver. The Auraria Campus, housing the University of Colorado Denver, the Metropolitan State University of Denver, and the Community College of Denver, lives on the same land. Several historic buildings from the gold rush era have been preserved on part of the campus and now house various departmental offices. They have some ghost stories of their own. As of this writing, Rocky Mountain Paranormal hasn't worked on those locations, but they remain on our ever-growing to-do list and therefore may be found in future volumes of this book series.

Regardless, the name stuck for the newly formed "Denver City" (though it was at that time more of a pass-through mining settlement than a true "city."

In other words, the 1851 Treaty of Fort Laramie was broken pretty much as soon as it was signed, and the land that would become Cheesman Park was right in the middle of it all. The consistent influx of people looking to make their fortune in the gold rush led eventually to the Treaty of Fort Wise in 1861 in which the Arapahoe and other tribes ceded further lands to the settlers. However, disputes would continue until the conflict came to a bloody head in the Sand Creek Massacre which took place near what is now Eads, Colorado in 1864. After that event, most of the Arapahoe and Cheyenne relocated.

In the meantime, the Colorado Territory was established in 1861, consisting in substantial part of the lands ceded by the indigenous tribes in the Treaty of Fort Wise, and would continue to exist as a territory until it was granted statehood in 1876, one hundred years after the nation's founding.

In order to oversimplify history a bit with respect to the Cheesman Park land specifically, it was either given to or stolen by (depending on how one wishes to read history) the newly formed Denver City from the Arapahoe, but it came with some strings attached. First, the Arapahoe told the City not to put a cemetery there because it was Arapahoe burial ground. Second, because they knew those kinds of conditions were seldom honored, especially in light of the fact that the City was already encroaching on their lands, they said, essentially, "if you must put a cemetery there, only bury people shallowly so you don't dig up any of ours."

Denver honored half of that agreement. They almost immediately began using the land as a cemetery, but buried the bodies approximately three feet below the surface (as opposed to the traditional six feet). Remember that fact for later. It will become relevant again.

In the winter of 1858 to 1859, William Larimer and William Clancey selected the land for a new cemetery. It was officially given its charter by the Kansas Territorial Legislature in 1860, but it was already in use. Remarkably, we actually know the identity of the first person buried there: one Jack O'Neill, who died in a gunfight, was buried on March 30, 1859. Another early "resident" in the cemetery was a certain Mr. John Stoefel who was hanged on a Cottonwood tree at the intersection of 10th Avenue and Cherry Creek Boulevard for killing his brother in law on or around April 10, 1859.

Importantly, this was not a tiny little local cemetery. This place was immense and consisted of several smaller "sub-cemeteries" grouped together in a single location. Mount Prospect, later called City Cemetery, was located where Cheesman Park is now, but the full cemetery's boundaries extended far beyond. On the north side of what is now the park a location was designated for the Grand Army of the Republic (GAR; a Civil War Union veterans society) graves. Just southeast of those (near where the pavilion is now) were the society graves. On the west side were burial grounds for

Freemasons and members of the Improved Order of Odd Fellows. A Chinese section and a potter's field were on the south side. Where the Denver Botanic Gardens now sits was once the Catholic Mount Calvary Cemetery. The Hebrew Burying and Prayer Ground was a bit further east, where the City of Denver maintains an underground drinking water storage area to this day.

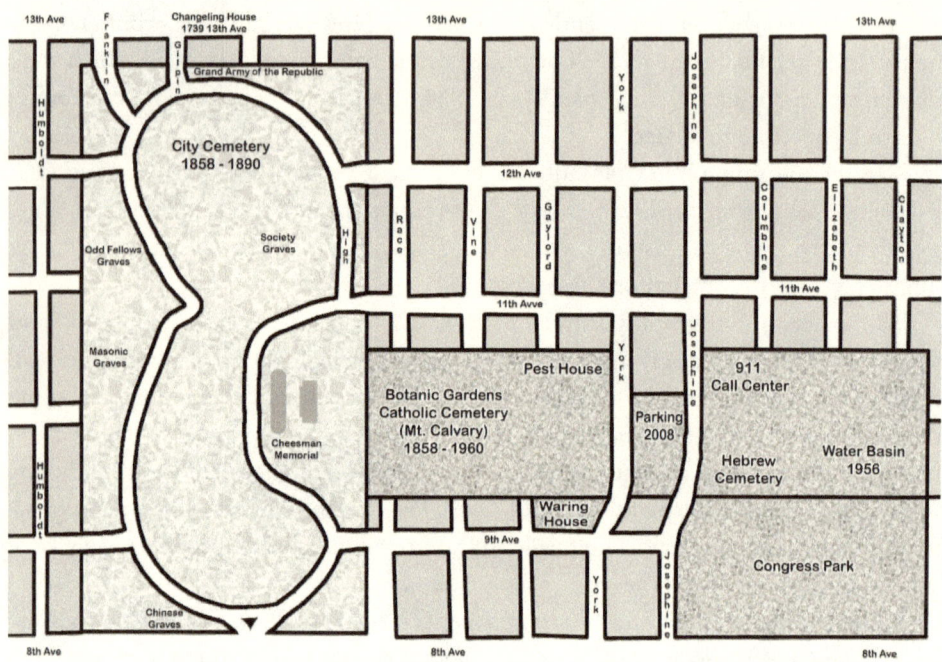

Figure 5.2. Cemetery locations in relationship to current structures. Image: Bryan Bonner.

To provide a bit of further background regarding the cemetery and an interesting historical note, the first undertaker at the cemetery was named John J. Walley. He was a cabinetmaker by trade and maintained business offices at 1412 Larimer Street (part of Larimer Square—see Chapter 2). In 1890, Mr. Walley along with Dr. A. M. Burknam and R. P. Rollins found themselves in a court of law, accused of "performing an unauthorized autopsy – Liability to heirs." They lost the case, which alleged that the defendants had performed an autopsy on one Tamar V. Thorpe on January 10, 1890 without the family's permission. According to the defendants, the body was given to them by a friend of the deceased with whom she'd been living. In order to have the body buried, they argued, there had to be a certificate of death and they had performed the autopsy toward that end. The trial court, finding the defendants' arguments persuasive, ordered the jury to find in favor of the defendants. Plaintiffs appealed, arguing this was an erroneous instruction, but P. J. Richmond, writing for the Colorado

Court of Appeals, affirmed the lower court's opinion.[2]

The potter's field is also of some historic interest. At the boundary of the cemetery there was a sort of hospital known as a "Pest House." These facilities, also known as plague houses or fever sheds, were used for patients with communicable diseases (most notably tuberculosis, cholera, smallpox, or typhus), often as a means of forced quarantine. Though the Pest House was located directly adjacent to the cemetery, the poor who had just died could not afford a proper burial, so their remains were dumped into a mass grave immediately behind the Pest House. This particular potter's field was located where the Denver Botanic Gardens now maintains the Community Gardens.

It's also worthy of note that not all of these cemeteries were founded or dissolved at the same time. The Catholic section (where the Botanic Gardens is now) was acquired by the Archdiocese of Denver under the direction of Bishop Joseph Projectus Machebeuf. This occurred at the time when the City Cemetery was being converted into a park, but Mayor Joseph E. Bates sold the 40-acre section to the diocese which was renamed Mount Calvary. Its graves were eventually relocated to Mount Olivet Cemetery and the land sold back to the city in 1950, making room for the Botanic Gardens.

The Hebrew section was purchased by the Hebrew Burial Society and designated the Hebrew Burying and Prayer Grounds in 1875. These bodies were not removed until 1923 when the land was permanently leased back to the city to become the underground drinking water storage area. One thing we've noticed when discussing the history of Cheesman Park and the surrounding areas during our public lectures is that many people find themselves slightly uneasy at the idea of drinking the former Hebrew cemetery. However, we are always quick to remind people that there is no cause for alarm.

With regard to the Chinese section: this portion of the cemetery was given to a large population of Chinese immigrants who lived in Denver's Chinatown, also known derogatorily as "Hop Alley" (the word "hop" referred to opium, and Hop Alley was notorious for its opium dens and brothels offering the services of "exotic" women). This district was approximately bounded by 15th, 20th, Market, and Wazee Streets. After the bodies from this section were removed, it was a shrub nursery until 1930.

It's important to realize that the way people interacted with death and cemeteries has changed throughout the years. Once upon a time, we had a much closer relationship with death. Maybe this was simply due to the fact that, until quite recently in human history, death was a fairly constant part of life. We now take it for granted that most people live to a ripe old age, that most children bury their parents rather than the other way around, and that none of us are in particular risk of an untimely death during the course of our daily affairs. At the time Mount Prospect/City Cemetery was operating, and through the vast majority of human history (indeed, until a time within living memory of some of our oldest people), this was not the case. Death was much

---

2   Cook v. Walley & Rollins, 1 Colo. App. 163 (1891).

more prominent in daily life.

One of the manifestations of these different attitudes toward death is how we view cemeteries. Today, we tend to view the cemetery as a place, hopefully far from home, where we put our dead to keep them out of sight and out of mind except on those specific days when we make a pilgrimage to remember our deceased. At the time Mr. Prospect Cemetery was in operation, this attitude would have seemed alien. In those days, the cemetery was meant to be a beautiful place and much more a part of daily life. Indeed, cemeteries served much the same function then as public parks do today. Families would bring picnics and spend an afternoon. Arguably, picnicking in a place that reminds you of your ancestors is a better way of spending one's time than merely picnicking in some nondescript grassy park, but that's an argument for another day.

The problem with the cemetery in Denver, though, was that this was hardly the beautiful park to which one might take one's visiting family members. In fact, it was falling into horrible disrepair. There are several reasons for this. Part of the blame probably falls on the shoulders of the original undertaker, John J. Walley, who did little to improve the property's condition. But we can't blame him entirely. Though Cheesman Park today is right in the heart of a bustling metropolis, Denver was a much smaller town at the time. The Cemetery was too far out of the way, so it was seldom visited and allowed to fall into disrepair.

Making matters worse, the rather informal acquisition of the land continued to raise some questions about the property's ownership. In 1872, the United States Federal Government determined that the land didn't actually belong to Denver. Rather, they claimed it for themselves as Federal land, arguing that it had been acquired not by the City of Denver but by the Federal Government through a land grant from the Arapahoe. This crisis was averted when the City purchased the land from the United States for $200 (no, that's not a typo—even adjusted for inflation, that seems like a bargain). In 1873, the name was formally changed from Mount Prospect to City Cemetery, though it's still more commonly referred to by the former name even today.

By the late 1880s, the city had had enough. The cemetery was in poor repair and falling into disuse. Real estate developers were beginning to eye the property as Denver expanded. Denver's leadership decided they wanted to turn it into a park. Colorado State Senator Henry Moore Teller persuaded the United States Congress to authorize conversion of the cemetery into a public park. In an Act of Congress dated January 25, 1890, this authorization was granted, and Senator Teller, in recognition of Congress's action, named the project "Congress Park." We'll get back to how "Congress Park" changed into "Cheesman Park" a bit later.

Once the wheels of change were in motion, Denver gave families ninety days to relocate their loved ones from City Cemetery to a new location. A few of the sections were spared from this problem. As mentioned above, the Catholic section was sold to the diocese and renamed Mount Calvary, though it, too, would eventually be closed and the bodies relocated. But for the majority of the sections, things weren't so easy.

Among the problems were the substantial potter's fields. Those people had no relatives to relocate their graves. And even if they did, enough of them were mass graves that sorting out who was who would be an impossible task. Making matters worse, even within some of the other sections, there simply weren't any relatives around to claim and relocate the bodies. At that time, Denver was more of a pass-through town. People stopped there for a while on their way somewhere else (typically further west). If they died, they'd be buried in the local cemetery, but the rest of their people would keep moving.

Therefore, though many of the locals did move their loved ones' graves, many more bodies had no one around to claim them. That posed a problem for the city's plans to quickly turn the cemetery into a park. Thousands of graves were left unclaimed. Record keeping at the time was not what it is today, and registration of vital records wasn't in general practice until around the mid-1920s. No one knew who many of these bodies were or where they should go.

In response, the city contracted a local undertaker named Edward P. McGovern to remove the bodies and relocate them to Riverside Cemetery.[3] For his task, he was to be paid $1.90 (about $63 in 2022 dollars) for each body relocated. His work began on March 14, 1893.

The work did not go as planned.

Recall what we said earlier: the bodies in this cemetery were not buried the traditional six feet under. Rather, they were about three feet below the surface. That may not seem like it would make much of a difference, but it turns out to make a *big* difference. The tradition of burying bodies six feet underground has its origin during the Black Death plague in Europe, probably as an attempt to avoid the spread of disease, which was ill-understood at the time. Most American cities and states don't actually require that standard, but it's a fairly common practice, partly as a matter of tradition, partly to deter would-be graverobbers, and partly as a matter of preservation. With modern burial techniques—including embalming, stronger caskets, and the use of concrete vaults—even minimal standards are sufficient. But if your burial involves no embalming, no vault, and a simple wood box, there's a much higher probability of microbes and soil erosion disrupting a grave at three feet than at six feet.

Undertaker McGovern quickly discovered this when he started his work. At about three feet underground, the caskets had completely decomposed. Rather than simply digging up the caskets and relocating them to the Riverside Cemetery, he needed to extract the bodies, place them into new caskets, and *then* rebury them. Several problems presented themselves immediately. First of all, given the size of City Cemetery, that meant thousands of new caskets would be needed. Unfortunately, at about this same

3   Riverside Cemetery, established in 1876, still exists. It is the oldest currently-operating cemetery in Colorado.

time, the city of Denver was experiencing a severe shortage of adult-sized coffins.

McGovern quickly depleted his supply of caskets and he could find no way to get a new supply. The City of Denver responded with a proposition: switch to children's caskets. It turns out, the city actually had a surplus of these smaller coffins at the time, so they could supply plenty of those. There was another problem. Children's caskets measured about 1 foot by about 3.5 feet. Adult corpses would never fit. The city told Undertaker McGovern essentially to do whatever he needed to do to make them fit and he'd be compensated for the extra work. He hired a crew of over a dozen people[4] to help with the ghastly task and set about breaking apart the bodies to fit them into the smaller caskets.

For this reason, McGovern's name is often uttered in hushed tones by those who know part of the Cheesman Park story. He's been remembered as "the evil undertaker" who chopped up bodies while moving the cemetery. However, we at Rocky Mountain Paranormal have been on something of a quest to restore his reputation. Despite the gruesome nature of his task, McGovern was only doing what he'd been hired and specifically instructed by the City of Denver to do. It was a dirty job, but someone had to do it. Further, if blame is to be assigned, it belongs surely to the bureaucrats in charge of the project, and not to the hard-working undertaker they employed.

Because McGovern's reputation has been so unfairly tarnished over the years, let's pause the story of City Cemetery and Cheesman Park for a moment just to provide a bit of his biographical background.

Undertaker Edward P. McGovern was born in 1851 and opened his office, "E. P. McGovern Undertakers," at the age of 27. It was located at 1442 Arapahoe Street. That building is no longer there, but the address is between what is now a convenience store and a pet grooming business. His business was open for 44 years and was quite successful. In addition to the gruesome work for which he's unfortunately remembered, he was also one of the founding members of the Colorado Funeral Directors Association in 1897. That organization maintains a strict code of ethics for its members.

In 1880, McGovern sold his business to W. P. Horan. This business would eventually become Horan & McConaty, which is still owned by the Horan family today and remains one of the region's largest funeral homes.

During his work at City Cemetery, the local residents were so grateful for his service—once again, recognizing that he was doing a dirty job that no one else wanted to do—that they voted him Grand Marshall of the Saint Patrick's Day parade.[5] Though

---

4   We've found conflicting reports regarding the actual number of crew members. Some say McGovern hired 18 people, and others say it was 13. We haven't found receipts to confirm the true number, though we prefer the latter account due to the number's delightful association with triskaidekaphobia (fear of the number 13).

5   Goodstein, P. (1996). *The Ghosts of Denver: Capitol Hill.* Denver, CO: New Social Publications. p. 286.

public opinion was not particularly fond of the cemetery relocation project as a whole, as we'll see, Mr. McGovern wasn't recast as "evil" until people retold the story out of its proper context.

In 1900, McGovern began referring to himself as "Coroner." His son Vincent McGovern joined the business as an assistant in 1903, followed by his other son Paul J. McGovern as Deputy Coroner in 1909.

In 1923, he opened a new funeral home at 620 East Colfax Avenue, not far from the state capitol building (which has some ghost stories of its own).

In 1925, E. P. McGovern died at the age of 75 years and is buried at Mount Olivet Cemetery. His name was replaced as owner of the mortuary by Martin J. McGovern. By all trustworthy accounts (read: those not just repeating ill-informed urban legends), Undertaker McGovern was an honest and hard-working man who is poorly remembered simply for his attempt to do a job no one else would have wanted to do.

McGovern's unfortunate task was not the only problem to befall the cemetery relocation project, however. Because this was not at that time a well-populated area (though it was beginning to become so more than in decades past), the project provided an irresistible temptation for graverobbers.

Graverobbing itself has an interesting history, which we don't have time to go into in its entirety. For our purposes here, it's worth noting that there have historically been two primary motivations for graverobbing. The one most people can think of on their own is graverobbing motivated by profit. Many people are buried with at least some of their jewelry or other valuables. With graves being opened and relocated anyway, this seemed like a perfect opportunity for some people to pick up some extra valuables without too much extra risk.

The second reason for graverobbing is to steal the bodies themselves. Primarily this was done in the past by "resurrectionists," graverobbers who exhumed recently deceased corpses and brought them to medical doctors or anatomists for dissection and study. Indeed, much of our early knowledge of humans' inner anatomy and physiology resulted from precisely this practice, since there was not yet a mechanism by which people could donate their bodies to scientific study, nor were there X-rays, MRIs, or other means to examine internal structures without cutting bodies open.

The time of the resurrectionists was nearly at an end by the time City Cemetery was in the process of becoming Cheesman Park. However, the practice had not entirely died out yet, and cadavers, no matter how illicitly obtained, could fetch substantial prices on the medical black market. It is not known whether any resurrectionists were active during the cemetery relocation project, but the possibility exists. At the very least, graverobbing in pursuit of jewelry and other valuable goods was in fairly wide practice at the time.

Indeed, there are even some allegations that some of the men Undertaker McGovern hired to assist with his work may have robbed some of the graves on which they were working. Importantly, these allegations are unsubstantiated and not direct-ed at McGovern himself, though some more recent publications have reported them as fact and, worse, attempted to shift the blame to McGovern.

Meanwhile, exhumations were also underway at the Chinese section. The Chi-nese community, unlike many of their neighbors in Denver, had maintained remarkably good records of their deaths and burials. Their goal was to reinter the bodies in their homeland, and the local Chinese community set about that work.

There are several important notes regarding this process. The first is that the local Chinese community did not consist of particularly wealthy people. Therefore, rath-er than hiring professional undertakers to perform the exhumations, they carried out the work themselves.

It's also necessary to remember that this predates Mao Zedong's seizure of China, so the local Chinese community primarily practiced traditional Han Chinese religion. Part of these religious practices involves ancestor veneration. One practice in particular is known as a "lucky burial," in which an individual's bones are dug up, washed clean, and stored for a time in a jar prior to reburial in a location determined to optimize the flow of qi. This was thought to help prevent spirits from haunting their relatives.[6]

Only the bones are necessary for this lucky burial, however. The Chinese didn't have any particular interest in the skin, meat, organs, etc. that were dug up along with the bones. So their idea was simply to scrape the bones clean and prepare them for the lucky burial, leaving everything else on site. Within the context of traditional Chi-nese religious observance, this makes perfect sense. To late-19th century Denverites, on the other hand, it must have presented quite a shocking sight.

An interesting side note regarding the Chinese grave relocation project: shipping bodies back to China was not an inexpensive proposition. Steamship companies at the time would charge full rates for the transportation of dead bodies. However, fish bones could be shipped by weight. Therefore, apparently the Chinese bones sent back to the homeland were placed in large boxes labeled "fishbone" and shipped from San Francisco to China at a rate of $20 per ton.

Not all of the problems facing the relocation project had their origin in the proj-ect itself or manifested at the same time. In a 1903 complaint concerning Mount Calvary (the Catholic section, which was not moved until some years later) published in the Denver Times, one Mr. Charles Cox explained that he'd seen how the paupers were buried in the potter's field sections of the cemetery. He claimed that the rotten remains of previous burials were being dug up and that sometimes two bodies were removed from the same grave and left out for hours behind the homes. The under-taker explained it was easier to dig in previously disturbed earth, and the new coffins

---

6   Berkson, M. (2016). *Death, Dying, and the Afterlife: Lessons from World Cultures Course Guidebook*. Chantilly, VA: The Great Courses. p. 130.

would lie flat against the coffin underneath.[7]

With all of these shenanigans going on, the whole thing became a political nightmare for the City. On March 19, 1893, the newspaper *The Denver Republican* ran an article about the relocation project under the headline "The Work of Ghouls," detailing the project's gruesome problems and blaming the paper's political rivals "Cap" Smith and John D. McGilvray.[8]

The term "ghoul" here may take on a triple meaning. In pre-Islamic Arabian lore, a ghoul is a monstrous or evil entity known to inhabit graveyards and consume human flesh. This supernatural entity has survived in modern supernatural lore, often referring to a kind of undead being. George A. Romero's classic zombie film *Night of the Living Dead* famously referred to its walking corpses as "ghouls," because the term "zombie," borrowed from Haitian voodoo lore, had not yet taken on its modern meaning of a mindless reanimated corpse. A second meaning comes from the aforementioned resurrectionists who would illegally exhume corpses to sell them to doctors or scientists engaged in anatomical study. Another term for these individuals, along with graverobbers, body snatchers, or resurrectionists, was "ghouls." Finally, through the first two meanings of the term, the word "ghoul" has taken on a meaning in common parlance to refer derogatorily to individuals who engage in dark or macabre work or pastimes. Arguably all three meanings could fit the newspaper's description of the happenings at City Cemetery.

While many of the newspaper's accusations are superficially true—bodies were indeed being broken apart to fit into small boxes—they conveniently ignored the context related to the casket shortage, the decomposition of the caskets in which individuals had originally been buried, and the available surplus of children's caskets. Indeed it is true that the division of bodies into multiple caskets resulted in higher fees paid to McGovern and his men, but that was negotiated as payment for the extra work they had to undertake in order to complete the now even more gruesome job.

Not yet satisfied with the uproar they had caused with the above article, the *Denver Republican* ran another article the following day, March 20, 1893, arguing that the new burial site at Riverside was not fit for use for its intended purpose. When the Platte River is flooded, the article claimed, the whole of the region might find itself underwater. While it is true that Riverside Cemetery was (and still is) bordered by the South Platte River on the northwest side, the cemetery management later denied the newspaper's allegations, claiming that the bodies were buried on the safer southeast side of the cemetery.

Regardless of the veracity of some of the claims' details or their lack of context, the whole operation had become a political nightmare. Denver Mayor Platt Rogers,

---

7   In: Arps, L. W. (1977). Cemetery to Conservatory, Part IV. *The Green Thumb*, *34*(3): 68-74.

8   "The Work of Ghouls." *The Denver Republican*, March 19, 1893.

recognizing that continuing the project could deepen the political crisis, ordered the project stopped. Graves that had already been exhumed were moved, and those that were still in place were sealed. The park would be built over any bodies that remained. Despite the *Denver Republican*'s assurances that this would be the correct solution to the problem, however, the story of Cheesman Park does not end there.

A couple of isolated incidents are worth mentioning here with regard to the Mount Calvary Catholic Cemetery, which remained active for a time following all of this trouble. Because this property was maintained as a cemetery until 1950, a fence was erected to separate the park from the cemetery. But because this particular plot of land seems destined never to have a cemetery operate without incident, by the time Mount Calvary was given back to the city, the fence had fallen apart in some areas and children commonly played in the cemetery. There's nothing necessarily wrong with that in and of itself, but there are unconfirmed stories that some of the children took to collecting bones and even some of the old headstones to use for their own purposes. Whatever those purposes might have been.

A morbidly and horribly humorous incident took place when Horace Tabor, the "Silver King," and one of the wealthiest Coloradoans of the time, died in 1899. It was such a monumental event that flags were flown at half-staff and it was reported that some 10,000 people attended his funeral. His body was interred at Mount Calvary. When that cemetery was eventually relocated, Tabor's grave was moved to Mount Olivet, where it now lies next to his second wife, Elizabeth "Baby Doe" McCourt Tabor (whose life inspired Douglas Moore's opera *The Ballad of Baby Doe*). When his body was exhumed, the city planned a grand celebration, even going as far as to hire a band for the occasion. However, the casket completely fell apart during the exhumation and the corpse fell out, at which time the family dog promptly jumped into the grave and ran off with one of Mr. Tabor's tibias.[9]

Returning to the transformation from City Cemetery into Cheesman park: for the moment, all seemed to be well. In 1898, architect and civil engineer Reinhard Scheutze completed the plan for the layout of what was then to be known as Congress Park (recall that Senator Teller chose the name in honor of the Act of Congress through which the City of Denver received authorization for the whole project). Because Mr. Scheutze died in 1910 before the park could be completed, S. R. DeBoer added the final parts of the plan. Things seemed to be moving in the right direction for Denver.

In 1894, still reeling from the aforementioned political crisis, the city closed off the land and erected a temporary wooden fence around the cemetery. The process of grading and leveling the ground began. But because the grave relocation project had been halted mid-operation, some of the open graves weren't filled in until as late as

---

9   Roberts, M. (2017). Threats to Historic Denver Cemeteries in Run-Up to Halloween. *Westword*. <https://www.westword.com/news/historic-denver-cemeteries-threatened-9633046>

1902. The rest of the bodies, whose graves were never opened, remained in place, with the park to be built directly over the remaining graves.

In 1907, the family of Denver's first water baron Walter Scott Cheesman, who had just passed away, were interested in finding a way to memorialize their loved one. Mr. Cheesman's wife Alice and daughter Gladys Cheesman-Evans approached the city with a proposal to name the park in Cheesman's honor. At the time, apparently, the powers that be in Denver were for some reason more fond of Congress than they were of Mr. Cheesman, so the initial proposal was rejected. However, when the Cheesmans offered $100,000 to rename the park and erect a pavilion in honor of Mr. Cheesman, the city acquiesced. A substantially smaller park to the southeast of what is now the Denver Botanic Gardens, is called Congress Park, perhaps a small token in remembrance of Senator Teller's original plan for the larger park.

Because the park is named in his honor, it's worth taking a moment to provide a few biographical notes concerning Mr. Cheesman.

Walter Scott Cheesman was born in Long Island, New York in 1838. He was the youngest of nine children of a prominent family. His father was involved in the honorable business of bookbinding and paper milling. Young Walter received private tutoring before entering the family business. As a young man, he lived in Chicago where he worked as a druggist with his brothers. In 1859, his elder brothers William Henry and Edward Talbot Cheesman came to Denver to open a pharmacy store. However, their business was struggling. Walter Scott followed his siblings to Denver in 1861 and took over the business. By this time, he already had a strong sense for business and was able to turn the store into a success. He maintained this operation until he sold it in 1874.

Meanwhile, Cheesman developed an interest in the railroad business. In 1868, he partnered with John Evans and David H. Moffat to build the Denver Pacific Railroad to Cheyenne, Wyoming. He served as president of this company for a period of several years and was also instrumental in the planning and construction of Denver's Union Station, the Denver Boulder Valley Railroad, and South Park Road, among several other railroad and transportation enterprises.

Indeed, his work in these and various other projects helped Denver become the major city it developed into. He was active in real estate investing and development, worked to establish financial institutions and mines, and was instrumental in securing a desirable location for Denver's courthouse. Without his real estate transactions, the city would have been forced to build their courthouse further from the bustling heart of the young city. Along with three other men, he founded what would eventually become the Denver Chamber of Commerce.

While working these various projects, he realized that water was going to be central to the success of Denver as a city. He became a principal investor and served as president of the Denver Union Water Company for several years. Through this company, he built systems of dams, reservoirs, and distribution systems to bring clean water to the city. One of his most important achievements was the installation of a series of irrigation ditches including Denver's City Ditch (which becomes relevant in some ghost stories that are planned for the second volume of this book series).

Outside of business, he was president of the Colorado Humane Society and was known regularly to come to the aid of children and animals in need. On one occasion, even after having become one of the wealthiest men in Colorado, he was observed wandering out in terrible weather to provide shelter to a horse that had been left in an open field. He was a staunch supporter of Judge Ben Lindsey's Juvenile Improvement Society.

Mr. Cheesman's personal life was not always happy. He married Bessie Lyster in 1869. Less than a year later, his wife died at the age of 24 years. The grief-stricken Cheesman would remain single for many years until, at the age of 47 in 1885, he married the widow Alice Eudocia Foster. They would have two children: Mason D. Cheesman, born in 1886, and Gladys Cheesman, born in 1887. Only Gladys would survive into adulthood. Young Mason died in 1900.

Walter Scott Cheesman died in 1907, prompting Alice and Gladys to offer their $100,000 donation to the city in exchange for memorializing Mr. Cheesman's life and work in the park.

Much like its cemeterial past, the park's future was destined for bureaucratic failures. The original design for the park (still called Congress Park in these early plans) called for such extravagances as a pond with water lilies on the west side of the park and importation of tree species that had "never been seen in the area." The Walter Scott Cheesman memorial pavilion was similarly planned to look much more spectacular than what eventually developed.

With regard to the latter point, the stone destined for the pavilion was quarried from the caverns of Treasury Mountain. The original pavilion included ornate stairways, gardens, and a complete lower level. What exists today consists only of the top half of the original structure.

Figure 5.3. The Walter Scott Cheesman Memorial Pavilion today. Photos: Bryan Bonner.

Construction on the pavilion was completed shortly after the park opened. However, it began falling apart almost immediately. On October 30, 1909, the City Council refused to accept the pavilion arguing that the cement work had been done improperly, the foundation was cracking, and there were concerns that the problems would only worsen with time. They blamed the Ladd Sanger Construction company, who had performed the cement work. The city refused to pay, so construction stopped.

The city's solution was as inelegant as one might expect from the bureaucrats involved in local government. They pushed up several tons of dirt around the building and buried the entire bottom floor of the pavilion, including its ornate stairways. Those who see the pavilion today (at least from a distance) might think it looks like a nice little structure, never understanding how much nicer it would, could, and should have been. But the problems don't end there. Closer inspection reveals even the part that's still standing has some structural issues of its own.

The pavilion's columns were never finished and polished. In addition to making this look inferior to what was planned, this has led to the slow degradation of the columns, which are beginning to crumble today.

Figure 5.4. The pavilion columns are unfinished and falling into disrepair. Photo: Bryan Bonner.

For reference, the Ancient Greeks constructed similar columns that have stood for thousands of years with little effort required to preserve them. It's disgraceful that the City of Denver couldn't finish a similar job and construct columns that would last more than a little over 100 years, especially after taking the Cheesman family's money to pay for the project. To date, there is no effort under way to fix the problems.

Several years after the city buried its failures, S. R. DeBoer was contracted to add the fountains currently seen on the pavilion's west side. In 1913, the Colorado Mountain Club donated the brass monument showing the mountain view, also on the pavilion's west side.

At this point things remained fairly constant for a time. The next big change was the aforementioned movement of the bodies from the Catholic Mount Calvary cemetery to Mount Olivet in 1950, making way for the installation of the Denver Botanic Gardens. There are some remarkable stories with regard to that operation as well, but we'll defer discussion of those until we get to discussing our own investigation in the interest of preserving a logical flow of information.

The important thing to realize about all of this history we've been discussing is that, though many people don't realize it, the bodies from City Cemetery are still there. Just three feet or so below the surface of a park where children and families picnic and play, right below the ground where people walk their dogs or enjoy their morning jog, lie the unmarked graves of some of Denver's early residents.

No one knows exactly how many bodies are still there. Estimates range from around 2,000 to around 5,000. Based on the research we've done, our best guess is that there are probably about 3,000 corpses still in Cheesman Park. And if that doesn't lend itself to a good ghost story, we don't know what does. Speaking of which….

Paranormal Claims

By rights, Cheesman Park should be one of Denver's most haunted locations. For that matter, it should be one of the most haunted places in the country. Supernatural lore often suggests that spirits haunt the world when their graves are disturbed. If that is the case, it seems like Cheesman Park should be ripe for somewhere between two and five *thousand* restless spirits.

However, the actual ghost stories that have been reported over the years are surprisingly few in number in light of the park's history and are often fairly nebulous and difficult to pin down. Nevertheless, we at Rocky Mountain Paranormal have done our best to catalogue the kinds of paranormal claims people have made over the years.

The police are remarkably good at keeping visitors out of the park after hours (defined by local ordinance, at least as of this writing, as 5:00 a.m. to 11:00 p.m.), but local legend maintains that if one visits the park on a certain moonlit night, all of the old grave outlines can be seen. Some think this is purely an optical phenomenon, while others think that it's a sign from the spirits trying to remind us of their presence just beneath the ground's surface.

One of the most unsettling claims, which has been reported by multiple witnesses, is that sometimes when one is reclining in the grass of Cheesman Park, one suddenly finds it very difficult to get up. It's as if someone or something is holding one down, pushing or pulling one toward the ground.

Others have reported the sound of whispering voices throughout the park, coming from unseen sources even when the witnesses were apparently alone.

Legend has it that there is a spot somewhere in the park—no one seems to know exactly which spot—where the sun never casts a shadow. Supposedly this has been covered up by the City of Denver. One variation of this story is that the spot in question is a particular tree—the Hangman's Tree—where an innocent man was hanged by an angry mob for someone else's crime.[10] Supposedly this tree never casts an afternoon shadow.

During the grave relocation project, according to the legend, one grave worker who had been looting from the graves claimed he was contacted by a ghost. His identity is not known.

Additionally, the park proper is surrounded on three sides by residential properties. Some rather magnificent historic houses and mansions border the property, and they have some ghost stories of their own, perhaps related to the park and perhaps independent thereof. One of those stories is that people have reported their homes being visited by mysterious strangers dressed in Victorian garb. While we have often been amused at the idea that Victorian attire seems to be a prerequisite for almost any ghost story, it actually makes sense in the context of Cheesman Park and the surrounding areas to the extent that, if the ghosts of Mount Prospect or City Cemetery are making house calls, they're likely to be the ghosts of people who died roughly during the Victorian era.

One house bordering the park is known as the Waring house. Located at 909 York Street, it's now part of the Denver Botanic Gardens and holds some of their administrative offices. It was built by architect Jacques Benedict for Richard and Margaret Patterson Campbell (who feature more prominently in stories surrounding the Croke-Patterson-Campbell mansion, which will be discussed in a future volume of this series). It's known as the Waring house because it was given to the Botanic Gardens by Dr. James and Ruth Waring in 1959.

Not uncommon for the time, the house featured some secret passages (all of us at Rocky Mountain Paranormal strongly encourage architects to build more houses with hidden rooms and secret passages—they lend themselves so well to a nice creepy story, and we imagine they're just plain fun to have in one's home). One secret door leads to a staircase that ascends to a bedroom. According to legend, one can lift the stairs to reveal another secret room. However, doing so angers the spirits who haunt the house. In his book *The Haunted Heart of Denver*, Kevin Pharris says workers in the house report that whenever the stairway passage is opened, the house becomes plagued by ghostly sounds and objects moving for a period of some weeks until the ghosts

---

10   It is known that one of the first people buried in the cemetery before it became Cheesman Park was indeed a hanged man. However, he's generally thought to have been guilty of the crime of which he was accused and was not hanged at the cemetery itself but at another location. It could be that his story *inspired* the urban legend, but the details don't match.

return to their state of rest.

While we're on the topic of the Botanic Gardens, allow us to relay a story that was sent to us by a reader of our website. Obviously this is merely an anecdotal story and we weren't able to investigate or corroborate it in any way, but it provides a good example of the kinds of stories reported in Cheesman Park and the surrounding areas, as well as an interesting peek into the kinds of things that show up in our mailbox. It is presented here mostly in the author's own words, but edited for grammar and clarity without changing the meaning:

> During the late 1970s, I lived in an old apartment building near Cheesman Park and the Denver Botanic Gardens. At the time, the Gardens did not have a set entrance fee, but were accessible in exchange for a donation. I went there often, usually in the late afternoons, remaining until closing. As a young woman alone, I did not like to walk in Cheesman Park (being more concerned about muggers than ghosts), but I felt safe in the Gardens.
>
> I had heard stories that the Gardens had at one time been part of a cemetery but did not think much about it until the year the Japanese garden was built. On a rather chill, rainy day in the fall of 1978, I went to see the progress on the new garden and saw a small marble child's headstone lying on a pile of dirt where the lake for the Japanese garden was being dug. There was a carved figure of a reclining lamb on top of the stone, freshly chipped from the digging. The name was too weathered to read, but the date of the death was in 188(?), aged three years and a few months (if I recall correctly).
>
> I'd been walking in the rose garden section of the Gardens earlier, as there were a few late roses still blooming, and had found a small white rose that someone, perhaps a child, had picked and thrown down. I'd picked it up and was carrying it. On an impulse, I laid it by the little headstone with a blessing and went home.
>
> I awoke the next morning, around dawn, to find a single, very fresh rose petal on my pillow a couple of inches from my face. It was a real, physical rose petal. It was not white, but a sort of deep pink color, and small, as if from an old-fashioned rose rather than a hybrid tea-rose. I picked it up and could feel and smell it. I said "thank you," and gently laid it on the night stand while I went into the bathroom and then to the kitchen to put the kettle on. When I came back to the bedroom, thinking I would press the petal in a book, the petal was gone. I would have wondered if perhaps my cat had taken it, but he'd been with me in the bathroom and then ran to the kitchen for breakfast and stayed there eating while I went back into the bedroom.

I could find no trace of the petal, but there was still a faint lingering scent of the rose on my pillow. I think perhaps someone left it for me in return for the white rose. As to why it should vanish once I'd seen it, I have no idea.

Another house bordering the park has a ghost story that completely revolutionized the horror genre. It is to that house that we now turn our attention.

In 1968, playwright and composer Russell Hunter rented the Treat Rogers Mansion at 1739 East 13th Avenue, just on the north side of Cheesman Park. Hunter had been a music arranger for the CBS affiliate in New York City but moved to Denver for two purposes. First, he intended to help his parents manage their Three Birches Lodge in Boulder. Second, he wanted a place where he could work on his music. He managed to rent the mansion for only $200 per month. Adjusted for inflation, that would be about $1,700 in 2022. Given the size and location of the mansion, that was a remarkably low price even in 1968. For reference, that inflation-adjusted amount is comparable to the price of rent for a single-bedroom *apartment* in 2022—never mind a mansion!

According to Hunter's testimony, though, he quickly discovered the reason for the incredible bargain: the house couldn't keep a tenant. Nobody wanted to live there.

Beginning soon after he took up residence in the mansion, Hunter began experiencing strange and troubling phenomena. He would hear "unbelievable banging and crashing" every morning at precisely 6:00 a.m., which would stop as soon as his feet touched the floor. Water faucets turned on and off by themselves. Doors opened and shut apparently of their own volition. Paintings flew off the walls. Even the walls themselves vibrated.

So frustrated with the sounds and disturbances, Mr. Hunter screamed "Stop it!" The noises abruptly stopped. But things wouldn't remain quiet for long, and the disturbances would resume regularly.

While this was all going on in the house, Mr. Hunter and an architect friend of his stumbled upon a hidden staircase in the back of a closet. According to Hunter's account, this stairway led to a hidden room on the mansion's third floor. Inside the room was a child's trunk. Inside the trunk were the journal and schoolbooks of an eight- or nine-year-old child (accounts vary as to the age) from a century prior. Said journal contained a dark and disturbing story: that of a young disabled boy who suffered a life hidden away in isolation. Within the journal's pages, Hunter also discovered the boy's favorite toy had been a simple red rubber ball.

A few days later, Hunter observed a red rubber ball bouncing down a spiral staircase in the house. Importantly, Hunter had no children of his own and lived alone. There was no one else in the house.

According to the story Hunter uncovered upon reading the journal, when the child, then given the name of Eric Evans, was born, he was set to become heir to a $70 million fortune on his twenty-first birthday. However, young Eric was a sickly child and not expected to attain that age. If the child died prior to attaining the age of majority (then defined as 21 years), the fortune would instead pass to a distant branch of the family. His parents, covetous of the money, could not countenance that idea and hatched a plan to ensure the fortune passed to the "correct" heirs.

They locked young Eric away in the attic and adopted a replacement son, similar enough in appearance that they could present the changeling child to the world as if he were Eric. The real Eric, meanwhile, had only his rubber ball as company or entertainment while alone in the attic.

Alas, Eric was not long for this world. Shortly after his solitary confinement began, he died. According to Hunter's story, Eric's body, along with a gold medallion, was unceremoniously dumped in a field on the southwest side of town.

When Hunter failed to act upon or go public with the information he'd discovered, the supernatural occurrences resumed and intensified. Unsure of what else to do about these mysterious goings-on, Hunter engaged the services of parapsychologist Robert Bradley and held a séance. During the séance, additional information was uncovered. Bradley informed Mr. Hunter that the spirit of Eric was not trying to threaten Mr. Hunter but was attempting to communicate his chilling story. Furthermore, the séance revealed that the replacement child had since graduated college and become successful in business. It also revealed the location of Eric's body. The field in which he'd been buried had since become the site of a home in South Denver. Finally, the séance made clear what Hunter's task was to be: find the home in question, dig below their closet to exhume the body, and use the gold medallion inscribed with Eric's birthdate as proof of the story. This would set the boy's spirit free.

According to the story Mr. Hunter told, he was eventually able to gain access to the home in question and to dig for the body. He claims he did find Eric's corpse, complete with the medallion. Once he took possession of the medallion, Hunter says that the walls of the mansion shook and the thumping sounds grew increasingly loud. He went as far as to claim that a glass door exploded and severed an artery in his arm, requiring medical attention.

After about a year's residence in the home, Hunter moved out and relocated to a home on Kearney street. A few months later, the Treat Rogers Mansion was demolished to make room for an apartment complex. But Hunter's story has one more surprise: when the wrecking crew arrived to demolish the building, he said, the mansion instead exploded, killing a worker. And even that wasn't quite enough of a climax. Hunter says the haunting followed him to his new home until he contacted a priest at the Epiphany Episcopal Church who performed an exorcism.

If the story we've just recounted sounds a bit familiar, it should. Except for the Denver location and the names of the players, the story we've just told is almost exactly the plot of the 1980 classic horror film *The Changeling* starring George C. Scott and directed by Peter Medak. If you haven't seen it, you really need to. It's not only one of the best ghost stories ever put on film, but one of the best *films* ever made, full stop. We can't say whether the film's genius lies more with Hunter's story, which formed its basis or with Medak's direction, but there's nothing like *The Changeling* to terrify grown adults of the sight of a simple red ball bouncing down a staircase.

Because Hunter's story is so extraordinary, we begin the discussion of our investigations into Cheesman Park with what we've been able to discover about what has become known as The Changeling House.

Our Investigation

The problems with investigating something like Cheesman Park are numerous. First of all, the park itself represents a large area. We're good at what we do, but we usually travel to locations with a fairly small crew. Covering so much ground is difficult. Worse, it's all outside, so it's incredibly hard to control environmental factors. Worse again, it's open to the public, making it all but impossible to control for all the possible confounding variables. And worst of all, the ghost stories, by and large, are vague and nebulous. It's difficult to pin down specific testable claims as we like to do at the beginning of an investigation. But when you've got a site with as much creepy history as Cheesman Park, we at Rocky Mountain Paranormal pay attention and work hard to find ways we can at least begin to scratch the surface of the paranormal claims.

A good place to start is with The Changeling House, also known as the Treat Rogers Mansion. Unfortunately the building no longer exists—in their infinite "wisdom," the powers that be seem to have decided we need another apartment complex (and worse: actually just a *parking lot* for the apartment complex) more than one of our magnificent historic mansions. But because the claims are specific and quite extraordinary (and because *The Changeling* is one of our favorite movies), we begin our investigation into Cheesman Park and the surrounding areas there.

We should begin simply by clarifying how the story became a movie. Russell Hunter, as a playwright and composer, was reasonably immersed in the entertainment business, so he knew how to get a story produced as a film. Indeed, Hunter is credited as the author of the story in the film's credits. The actual screenplay was written by William Gray and Diana Maddox. But since Hunter was involved in the production on at least some level, the film's details are remarkably faithful to the story Hunter told (and maintained was true).

The title itself comes from European (and particularly Irish) folklore, in which a changeling is a human-like creature—often a fairy—that has taken the place of a human child stolen by other fairies. Believers in the folklore lived in fear that their own children might be replaced by a changeling. However, if they discovered the impostor and threw the changeling into a fire, it would disappear up the chimney and bring their own child back. Some folklore alludes to horrifying stories of children being burned to death in this manner by paranoid or delusional parents.

Obviously, Hunter's story has nothing to do with fairies, but because the sickly Eric was replaced by an impostor—a sort of changeling—the title is quite appropriate. Arguably, much of the film's success stems from its masterful combination of supernatural horror with human horror—the story of how young Eric was isolated and replaced by a changeling impostor to guarantee the wealth of his parents.

Though the movie is a work of near perfection, the allegedly true story has some problems. For one thing, Hunter's story constantly shifted. Even such important details as Eric's age changed from telling to telling, often oscillating between eight and nine years but sometimes even going as far as twelve or thirteen years. That doesn't discredit the story, of course, but it does raise some questions about the source's credibility.

We do know some of the history of the mansion itself which we omitted from the history section of this chapter, preferring to place it here for the sake of logical continuity. That history further calls Hunter's story into question. Writing in *The Ghosts of Denver: Capitol Hill*, local historian Phil Goodstein explains that the house was built by the childless (but generous toward children) Henry Treat Rogers and wife Kate, after whose deaths the house passed to their niece, one Frances Ristine, also childless.[11]

Immediately we begin to see the problems with Hunter's story. There never was an Evans family living in the house about a century prior to Hunter's residence because the house itself did not exist 100 years prior to Hunter's residence. It was built some 76 years before his story took place. Furthermore, the initial residents—Henry and Kate Rogers—inhabited it until 1931, and their niece until 1934. There is no record of any Evans family ever living in the house.

Similarly, there is no record of any Evans family possessing the kind of fortune spoken of in Hunter's story. A $70 million fortune would have been unheard-of in 19th century Colorado (or even in early 20th century Colorado). Even if someone had somehow amassed such a fortune, they would not have gone unnoticed by the business and society pages of the newspapers, and there would be a substantial record of their activities. That's not to say it would have been impossible to sneak a changeling into their son's place without being caught, but at the very least the world would be aware of their and their son's existence. No such records exist.

In case we think perhaps Hunter merely got the name wrong, we hasten to point out that there is no parallel between the Rogers' story and the alleged Evans story. The Rogers were childless. And while they were far from poor, they never had the kind of fortune spoken of in Hunter's tale. There's no reason to think Henry and Kate Rogers were ever anything but wonderful and successful pillars of the community. As much as we love a good ghost story—and we most certainly do—it's unfortunate that when one tries to find biographical information concerning Henry Treat Rogers or his family,

---

11   Goodstein, P. (1996). *The Ghosts of Denver: Capitol Hill.* Denver, CO: New Social Publications. p. 309.

one's web searches are overwhelmingly dominated not by his actual life and accomplishments but by retellings of the Hunter ghost story.

Hunter's other stories related to the house are equally dubious. For example, he claimed to have actually discovered Eric's body (and the accompanying gold medallion) beneath a nearby home. The most obvious objection to this story is that despite years of searching, neither we nor anyone we're aware of has ever found a police record or newspaper article describing the discovery. Most unusual, as that sort of discovery would undoubtedly make for an enormous local—if not national or international—scandal. Even the alleged method of discovery seems implausible. We don't know about the rest of you, but if some strange man came to one of our houses and wanted to dig up the floor to search for a body, we're pretty sure we'd have him removed rather than give him permission to do so. And we say that as people who are typically keen on almost any kind of unusual adventure.

Then there's the matter of the house itself exploding upon the time of its scheduled demolition. Once again, this is the kind of happening that would assuredly make the local papers. The record remains silent on the issue. We have been unable to locate any record of a construction worker being killed—by any means—during the building's demolition nor any record of an unexpected explosion. While we always have to remain cognizant of the fact that absence of evidence is not necessarily evidence of absence, we also recognize that the absence of evidence *where one would expect to find it* does tip the scales a bit toward evidence of absence. It is our opinion that the alleged explosion never occurred.

Finally, and perhaps worst of all, researchers at the Denver Public Library could not even confirm that Russell Hunter had ever rented the property.[12] Despite the lack of concrete proof, though, we're prepared to take him at his word that he did rent the mansion, but we remain skeptical of all the rest of his claims.

Most of the remaining ghost stories resist forensic examination because of the size of the park and the irregularity of their occurrences, but we'll nevertheless report what we've been able to discover so far before we close this chapter with what ended up becoming one of our favorite investigations in our entire history.

With regard to the story that grave outlines can be seen on a certain moonlit night, a variation of the story turns out to be true, though not for any supernatural reason. We've never been able to see the outlines on a moonlit night (though not for want of trying), but we've discovered that under certain conditions, grave outlines can indeed be seen. If you want to see them, you'll have to wake up early in the morning following a light snowstorm. When the morning sunlight strikes the ground of Cheesman Park at just the right angle and with just the right dusting of snow, you can indeed see

---

12 Rudolph, K. (2013). The History of the Denver House That Inspired a Horror Film. *Denver Public Library: Genealogy, African American, & Western History Resources.* <https://history.denverlibrary.org/news/history-denver-house-inspired-horror-film>

the outlines of where at least some of the graves once were.

This is because there are ever-so-slight indentations in the ground in those locations. Because the graves were shallow and the caskets all disintegrated and collapsed, there are slight recesses in the ground throughout the park. These little dips are so slight that under normal conditions, they're essentially impossible to see. But the slight dusting of snow combined with the sharp angle of the morning sunlight renders them visible. Whether a similar phenomenon can occur on a moonlit evening remains unknown to us, but we haven't yet been able to recreate it.

One thing we've also discovered over the years is that because the graves are difficult to see with the naked eye, but it is well known that there are graves under the park, Cheesman Park has become a training ground for certain kinds of forensic investigators. There is a 501(c)3 non-profit organization called NecroSearch International. Their mission, since 1988, is to assist law enforcement with the location of "clandestine graves." That is, they employ a multidisciplinary approach to locating bodies in unknown locations, including missing persons and murder victims. Because they know that there are bodies still buried under Cheesman Park, they use the park as one of their training sites. For instance, they may use the park to teach their members how to use ground-penetrating radar to locate unmarked graves.

The story that ghosts or other entities hold people down in Cheesman Park is difficult to investigate. It's been reported by more than one witness, but it happens sporadically and is impossible to measure. All we can really say conclusively is that our members have spent plenty of hours reclining in Cheesman Park hoping to feel the touch of a ghost but as of yet none of us have replicated the experience.

Because the stories of the spot where the sun casts no shadow vary with respect to the spot's precise location, we've had some difficulty in thinking of a good way to investigate it. However, the variation involving the so-called Hangman's Tree seemed like the sort of thing we could look into. Unfortunately, we had no idea which tree was meant to be the one in question. Not relishing the prospect of individually photographing every tree in the park on sunny days to see whether they cast shadows (though we would have done it if we had to), we thought of an alternative idea. One of our members called up a satellite image of the park on the Internet and we began a process of exhaustively verifying that every tree in the park did indeed cast a shadow.

Angles of photography and clustering of trees made identification of some of the shadows difficult. However, after many hours (spread over a few days) of staring at aerial photographs of trees, we were able to determine that, at least as of the time at which the satellite photographs were taken, every single tree in or bordering Cheesman Park did cast a shadow. Paranormal investigation is glamorous business, indeed.

With regard to the various stories of people seeing ghostly individuals in Victorian attire, hearing disembodied voices, or feeling cold spots in or around Cheesman Park, there's little we can do. We can say that in the time we've spent in the Park (which has been a substantial amount of time), none of us have ever experienced any of these phenomena. But as of yet, we can't find any way of performing a forensic investigation, so we have to either believe or disbelieve those stories on their own, without the benefit of any more in-depth investigation.

Recall that in the late 1950s, the Warings gave a house, now known as the Waring House, to the Botanic Gardens. It's located on the southeast corner of the Botanic Gardens' property, at the intersection of 9th Avenue and York Street, and is used by the Gardens as an office space. In the 1980s, they discovered the foundation of the house was shifting. In response, they called an engineering firm to see what could be done to stabilize the building.

The firm came out and took core samples of the support dirt under the foundation, only to discover they'd accidentally core-sampled a casket (complete with skeleton) which appeared to be in a *vertical* position about twelve feet under the building.

Admittedly, the mere discovery of a casket under the building is hardly evidence for or against any ghost story. Nevertheless, when something weird happens, Rocky Mountain Paranormal takes notice. Even though all of the Mount Calvary graves had supposedly been relocated to Mount Olivet before the land was given to the Botanic Gardens (unlike those in Cheesman Park proper), it was unsurprising to us that some got left behind. However, given that most (if not all) of the burials in the old cemetery were shallow, we wanted to know how the casket got twelve feet under the surface. And we *really* wanted to know why it was vertical instead of horizontal.

One of the most important lessons we can impart with regard to paranormal investigation is that you never know what kind of expert you're going to need to call in to consult on a case. In this case, we reached out to a geologist and asked "what can you tell us about the dirt under the Botanic Gardens?"

One question you never want to ask a geologist unless you're prepared for a long dissertation is "what can you tell us about dirt?" Nevertheless, it was a fascinating conversation. We learned more about dirt than any of us probably ever wanted to know, but we were also able to come up with a hypothesis concerning the strange discovery at the Waring House: the dirt contains bentonite.

If you're a builder (or perhaps even a homeowner) in Denver, you may very well know that bentonite is the enemy. Bentonite is an absorbent and swelling claystone consisting mostly of montmorillonite. That probably doesn't mean much to you unless you happen to be a geologist but to put it into layman's terms, bentonite's absorbent properties are such that it expands substantially when wet. This makes it quite useful in many industries, but it makes it a nightmare for construction. When wet, not only does it expand, but it becomes almost fluid in nature. If you build the foundation of a building on bentonite, you can expect a lot of things to move around.

Unfortunately, much of Colorado's front range contains a lot of bentonite in the soil. For example, there's a dreadful section of E-470 south of Denver International

Airport where drivers notice the road suddenly becomes horribly bumpy. Many of us have been quick to blame improper investments into infrastructure projects for the condition of our roads (and sometimes that's the correct conclusion), but in the case of E-470 it turns out bentonite is to blame. The wetted bentonite under the road expands and exerts enough force to shift the road itself. There have even been stories of buildings built on bentonite splitting in half due to the shifting foundations.[13]

It turns out that's the most likely explanation both for the foundation trouble at the Waring house and for the unexpected discovery of a vertical casket. While it is possible to build on bentonite, it requires special preparation to secure the foundation and allow for the clay's swelling behavior. Old buildings such as the Waring House would have been built without knowledge of these techniques. And graveyards…well, there aren't many techniques to bentonite-proof a hole in the ground.

The result of all of this is that not only have some caskets become vertical, but some of the graves have migrated, entirely underground, several blocks outside of the boundaries of the original cemetery. Ghosts or no ghosts, there's some spooky geology at work in Cheesman Park.

On one occasion, we told this story during one of our public lectures. A woman sitting in the back of the crowd spoke up.

"That explains it!" she exclaimed.

Our antennae immediately perked up. When someone hears a story about migrating corpses and says "that explains it," we know there's a story we need to hear about. So we approached her after the lecture and asked her to elaborate.

She explained that she lived to the west of Cheesman Park. Her home was near, but not within, the original cemetery boundaries. One day the family dog had been playing in the back yard. When it came back inside it had in its mouth what appeared to be a human pelvis. Alarmed by the dog's discovery, our audience member called the police. Officers arrived shortly to collect the pelvis, said "thank you," and left without another word. She was never able to get any follow-up information until our story finally connected the dots.

Returning to the Waring House, we've never conducted a proper paranormal investigation of the building's interior[14], so we can't say anything about the couple of ghost stories they have there. But how cool is it that we were able to bring a geologist in on a paranormal investigation and actually figure out why there was a vertical casket twelve feet under the Botanic Gardens' administrative offices? These kinds of discoveries are part of what keep us going when the going gets tough.

But that's not the end of the story with regard to the Botanic Gardens. In 2008, construction began on a new three-floor parking structure on the east side of the

13   Williams, J. (2018). The Dangers of Building on Bentonite. *Colorado Builder.* <https://coloradobuildermag.com/build/the-dangers-of-building-on-bentonite-clay/>

14   It's on our to-do list, assuming the proper arrangements can be made.

Gardens, between York and Josephine Streets. This was a worthy project. Anyone who lives in or around Denver (or probably any medium-to-large city) knows that parking is an endless nightmare, so if parking can be added without destroying a historic building, we at Rocky Mountain Paranormal are all for it.

But the construction did not go as planned.

In November of 2008, construction came to a halt when crews digging for the new structure came upon what they thought might be human remains. They'd known about the location's history, so they didn't panic. Instead they stopped their work and contacted the Medical Examiner's office, who assisted the Botanic Gardens in the removal of the unearthed bodies, which were eventually reburied at Mount Olivet Cemetery, as the original graves relocated from Mount Calvary had been.

Because this was not an investigative operation on the part of the Medical Examiner's office, there was no autopsy report (and, in fact, no autopsy performed), so little information was discovered about the bodies. However, the M.E.'s office was able to confirm they had discovered and relocated "approximately" fifty-four bodies.

The reason for the imprecise number is that the bodies were first discovered with industrial construction equipment. Once remains had been found, work switched to hand tools, but between the construction equipment in 2008 and the bentonite for decades before, many of the bodies had been scrambled and scattered. The Medical Examiner's office reported that the pieces they recovered were most consistent with 54 bodies, but the precise number could never be ascertained.

The fact that so many bodies were being unearthed and reburied over the years without their identities being known was troublesome to us. One of the things we at RMP pride ourselves on is an unwavering respect for both our history and for our dead. We wanted to find out anything we could about the people involved in these stories. Unfortunately, as we already mentioned, there were essentially no records for the people buried in City Cemetery (particularly in the potter's fields) as death records weren't required until 1900 nor regularly maintained until around 1925 in Denver.

However, since Mount Calvary had been the Catholic cemetery, we had some hope that perhaps the Catholics would know something about at least their own people. Churches are remarkably good at keeping records. An argument could be made that the Mormons are the champions at keeping records, but if that is the case, the Catholics must be a very close second place.

We contacted the Archdiocese of Denver and asked what they knew about the people who'd been buried in Mount Calvary. They told us there were over 20,000 bodies buried there during the time it was an active cemetery. They only had records for six of them.

Those people (and the years of their deaths) were:
- July 10, 1864: The child of James Clifford,
- 1865: James O'Haire (3 years old) and James O'Haire (28 years

old),

- 1865: F. J. Smith (4 years old),
- 1866: Ellen Howlett (22 years old), and
- July 26, 1866: James Clifford (28 years old).

As glad as we are to be able to put names to at least some of these individuals, let this be a lesson to us all. Whatever our business, we need to keep good records. It's not right that a life should pass unremembered.

Our final and most substantial work with regard to Cheesman Park began in October of 2010 when an irrigation project was underway at Cheesman Park.

Knowing what's buried under the park, the City doesn't particularly like to dig there, for obvious reasons. But when you want to put in a new sprinkler system, it's going to involve some digging. During this process, several more bodies were unearthed. The way they were discovered and the way Rocky Mountain Paranormal got involved is a rather amusing story in its own right.

One of the irrigation workers was going about his business, tending to his work, when his digging unearthed some skeletal remains. Presumably he was aware of the park's history because instead of calling the police, he casually placed the bones into a box or bag marked "sprinkler parts" and went back to work. There are two variations of the story that follows, and we've been unable to conclusively determine which was the true course of events.

The more amusing version of the story is that the following morning, someone jogging or biking through the park happened across this bag marked "sprinkler parts." Either he took it home with him in the hopes of getting some free sprinkler parts or investigated on site and discovered the bones. Either way, as anyone might do, he ended up with the bones and the sprinkler parts at his home. Immediately upon making the morbid discovery, he did what any fine upstanding citizen would do. He dropped everything he was doing, picked up the phone and dialed…the KBPI radio morning show.

The alternative story, somewhat less amusing, is that a park ranger discovered the parcel containing the bones and reported it to the proper authorities. Meanwhile, the irrigation worker himself anonymously contacted the KBPI morning show.

Either way, the cat was out of the bag, so to speak. While the City of Denver is usually interested in keeping these discoveries quiet, the feature on the morning radio meant people knew human remains had been found in Cheesman Park. Most importantly, word reached our ears that another set of bodies had been discovered. It was initially reported that four bodies had been extracted, though it was later amended to five bodies, as two of them had become mixed.

As we've said before, when the weird happens, Rocky Mountain Paranormal is prepared to act. We immediately picked up the phone to contact the Coroner's office (we were later informed she strongly prefers the term "medical examiner"[15]). The conversation went something like this:

Medical examiner: *answers phone*.

RMP: Hello, I'm with Rocky Mountain Paranormal and I understand you've found some bodies in Cheesman Park, and—

Medical examiner: *hangs up*.

RMP: *calls back* Don't hang up! Just hear me out. Yes, I am with Rocky Mountain Paranormal. However, what I'd like to propose is to assemble a team of forensic anthropologists from the local university to examine the bodies and see what we can learn about them, and we'd like to document the process.

She fell for it! And so we contacted the Human Identification Lab, a forensic anthropology program at the Metropolitan State College of Denver (later renamed the Metropolitan State University of Denver). A team was assembled consisting of the Medical Examiner's Office, a forensic anthropologist and her students from the Human Identification Lab, and the Rocky Mountain Paranormal Research Society. Working in the local morgue (with an autopsy being performed in the next room, no less, as one of our weaker-stomached members discovered to his personal horror), we set about examining the bodies to see what we could learn from them.

The initial examination revealed the possible presence of one female and three males. Other than bones, the only other items present were handles from the coffins, nails, a hair comb, a single bone clothes button, and a .22 caliber bullet casing.

At one point during the examination of one of the skulls, one of the anthropology students became very excited. When asked what all the fuss was about, she pointed to a substance inside the skull and said "brain lining." If you're not accustomed to the behavior of scientists, this may seem strange. However, when one spends one's life

---

15   Though the terms refer to similar positions and are often used interchangeably, we learned through this interaction that there is actually a difference between the two jobs. A coroner is an elected official who may or may not be a medical professional. In the case when one is not a medical professional, they must contract forensic pathologists to perform medical examinations of the deceased before signing the relevant documents. In that sense, the coroner is a legal professional who works closely with medical professionals. By contrast, a medical examiner is a medical doctor who is board certified in forensic pathology, and is not an elected position. Educational requirements for the job of medical examiner are much more strenuous than for the job of coroner (which probably explains why the medical examiner in this case had such strong feelings on the topic).

Jurisdictions in the United States are divided. Some have coroner systems in which the coroner makes legal rulings on the basis of medical information provided by a hired forensic pathology staff. Others have a medical examiner system in which the medical and legal decisions are made by the same person.

dedicated to the study of a scientific discipline, the discovery of anything unexpected related to one's work is cause for substantial excitement. It was the only soft tissue discovered during the examination.

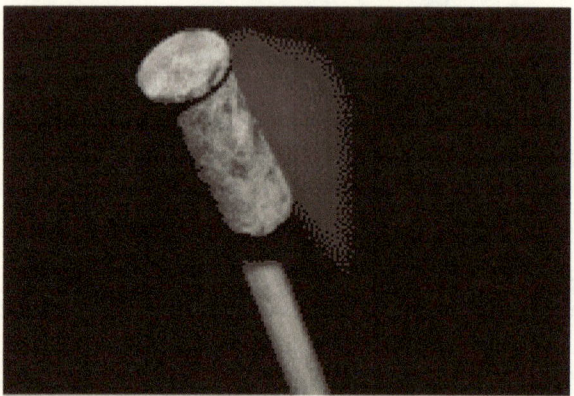

Figure 5.5. Bullet casing (.22 caliber) recovered from Cheesman Park. Photo: Bryan Bonner.

The reason there was no soft tissue (other than the one piece of brain lining), clothing (other than one piece of cloth casket lining), or wood from the coffins or caskets has to do with the shallow burials (recall that burials in this cemetery were shallow to avoid interference with the Arapahoe burial ground beneath). At three to four feet beneath the surface, the bodies would have been exposed to cycles of wet and dry as the seasons changed, resulting in harsh and acidic conditions that would have accelerated natural decomposition, not only of the human soft tissues but also of the wood caskets themselves.

More detailed examination followed, during which it was discovered that two of the bodies had been mixed together, so there were actually a total of five skeletons (or partial skeletons) present. During this more detailed exam, the skeletons were cleaned, measured, and photographed to document them. They were examined for bone structure, injuries, or any other clues that might provide information about their lives or identities.

Though of course the team wasn't able to provide specific names for any of the bodies, several important discoveries were made.

One male skeleton was determined to have belonged to a rather large black man the anthropologists believe was likely a stage coach driver. They were able to make this determination on the basis partly of the nature of injuries he'd received during his life, but also because the pattern of wear on his teeth indicated that he'd been eating rather well. At the time the cemetery was active, the most likely way for a black man to have been able to afford to eat well would have been if he'd been a stage coach driver.

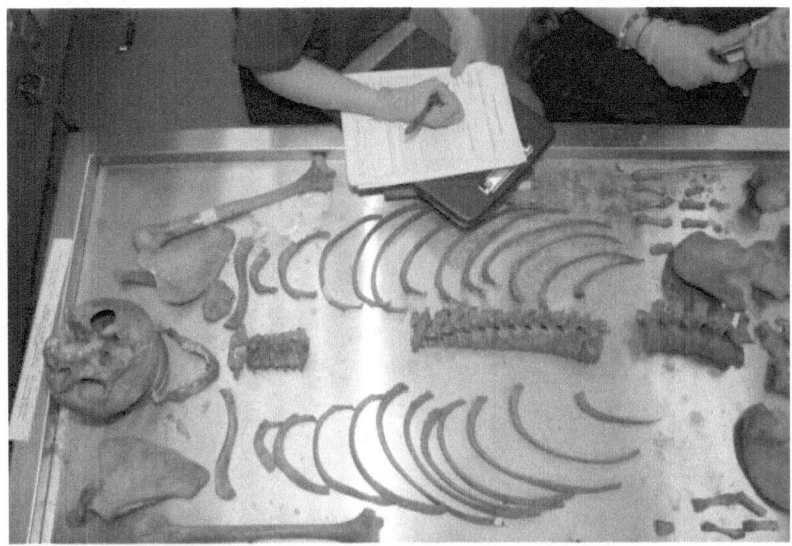

Figure 5.6. Skeleton of stage coach driver. Photo: Bryan Bonner.

Another body was determined to have been a young woman, around 14 to 16 years of age, of mixed Caucasian and Chinese descent. She would have had a very hard life. Being of mixed race, at that time, neither side of her family would likely have accepted her. Her remains showed signs of a short and difficult life marked and ultimately cut short by illness.

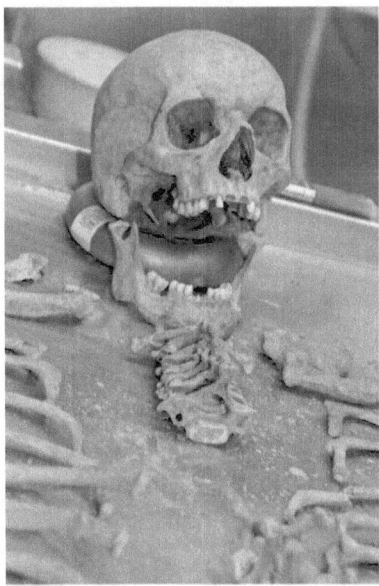

Figure 5.7. Skeleton of half-Chinese girl. Photo: Bryan Bonner.

Little could be determined about the other three skeletons except that one of them was male.

Participating in scientific research is always a pleasure. Several of our members have different varieties of scientific backgrounds, and all of us are dedicated to the scientific study of claimed paranormal phenomena. Even still, it's rare to be able to participate in something as remarkable as the forensic examination of multiple human bodies. It's a humbling experience, and it's one of the things we're most proud of in all of our case files.

As an addendum to the story, after the examination was complete and the bodies had been reburied, we received a phone call from the Medical Examiner's Office. The various non-human pieces recovered from Cheesman Park (primarily nails, handles, and other casket hardware, along with the one piece of cloth casket lining) were not reburied with the human remains. The Medical Examiner explained that she'd contacted everyone she could think of—the local museums, historic societies, and so forth—and none of them wanted the pieces. She asked if we wanted to take them off her hands.

Of course we immediately agreed. In addition to our work in forensic paranormal investigation, our members are also complete history nerds and collectors of historic and macabre artifacts. Within our homes, we maintain miniature museums of strange artifacts. We now house and care for the casket hardware removed from Cheesman Park. At the risk of sounding self-aggrandizing, we don't think anyone else would do a better job of caring for these artifacts or ensuring they're remembered and studied by the interested public.

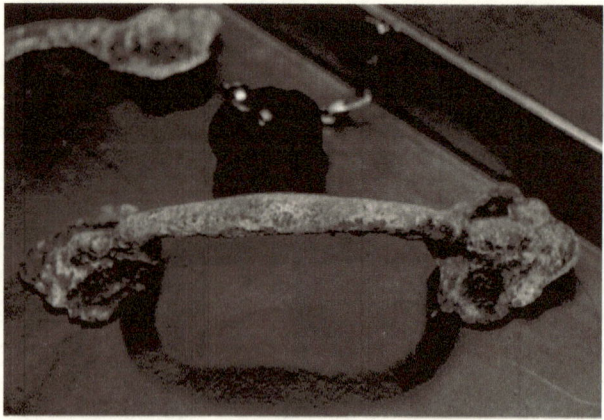

Figure 5.8. Casket hardware recovered from Cheesman Park. Photo: Bryan Bonner.

One of the questions we're often asked when we give public lectures is whether we ever speak to the dead. Honestly, we speak to the dead all the time. But there's an important follow-up question: have the dead ever spoken back?

We can honestly say: just once.

The one time we spoke to the dead and heard their reply was around the examination tables in the Medical Examiner's office when we asked five of the dead to tell us their stories and some of them answered. As we said before, it's not right for a life to pass unremembered. One of the things we're most proud of in all the work we've ever done was this one time when we were able listen to some stories from the dead and share them with the rest of the world.

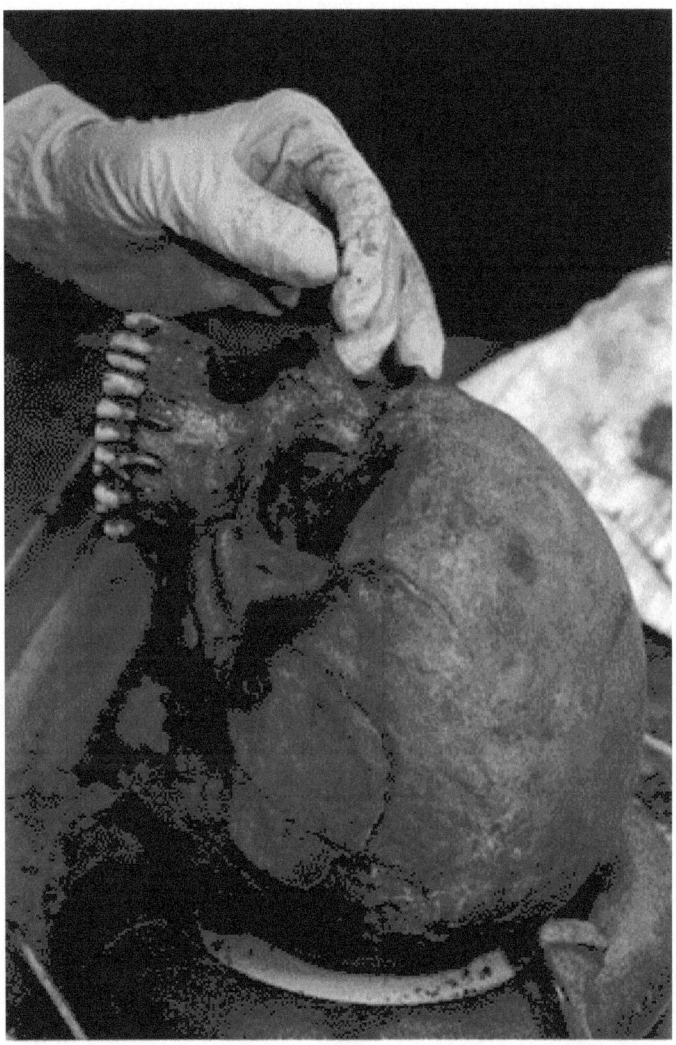

Figure 5.9. The dead tell their stories. A researcher examines a skull recovered from Cheesman Park. Photo: Bryan Bonner.

References and Further Reading

Arps, L. W. (1977). Cemetery to Conservatory, Part IV. *The Green Thumb, 34*(3): 68-74.

Berkson, M. (2016). *Death, Dying, and the Afterlife: Lessons from World Cultures Course Guidebook.* Chantilly, VA: The Great Courses. p. 130.

Getz, C. O. (2010). *Weird Colorado: Your Travel Guide to Colorado's Local Legends and Best Kept Secrets.* New York: Sterling.

Goodstein, P. (1996). *The Ghosts of Denver: Capitol Hill.* Denver, CO: New Social Publications.

Lamb, K. (2016). *Ghosthunting Colorado.* Clovington, KY: Clerisy Press.

Pharris, K. (2011). *The Haunted Heart of Denver.* Charleston, SC: Haunted America, The History Press.

Roberts, M. (2017). Threats to Historic Denver Cemeteries in Run-Up to Halloween. *Westword.* <https://www.westword.com/news/historic-denver-cemeteries-threatened-9633046>

Rudolph, K. (2013). The History of the Denver House That Inspired a Horror Film. *Denver Public Library: Genealogy, African American, & Western History Resources.* <https://history.denverlibrary.org/news/history-denver-house-inspired-horror-film>

"The Work of Ghouls." *The Denver Republican*, March 19, 1893.

Williams, J. (2018). The Dangers of Building on Bentonite. *Colorado Builder.* <https://coloradobuildermag.com/build/the-dangers-of-building-on-bentonite-clay/>

6
Into the Darkness:
Cave of the Winds

Colorado's mountains are full of caves and mines. Probably the most famous of the former is the Cave of the Winds, located in Manitou Springs, just west of Colorado Springs near the Manitou Cliff Dwellings. Its fame is due both to its history and its continued operation as a tourist destination open to the public. It also has some remarkable ghost stories, so Rocky Mountain Paranormal was excited to check it out. Paranormal investigation deep inside a cave, however, presented some of the most challenging conditions we've ever experienced, making this one of our more unusual case files.

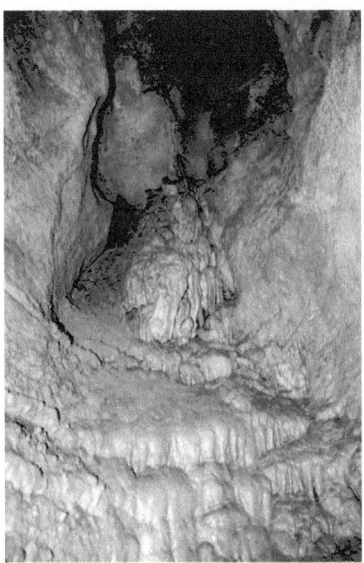

Figure 6.1. Cave formations in Cave of the Winds. Photo: Bryan Bonner.

The History

Because Cave of the Winds is a natural structure rather than a manmade building, it's necessary to discuss both the natural and the human history of the cave and its surrounding areas.

Our tale begins some 500 million years ago during the Ordovician Period, part of the Paleozoic Era. Immediately following the Cambrian Period with its famous "explosion" of biodiversity, the Ordovician was marked by continued evolution of new species. Invertebrates ruled the seas, though early fish—the very first vertebrates—are thought to have evolved during this period.[1] Also during this period, the Earth saw its first land flora.

During this period, the Pikes Peak region of Colorado, which includes the Cave of the Winds, was covered by warm and shallow seas filled with invertebrate life. Trilobites in particular were common. Fossil hunters to this day can often find trilobite or gastropod fossils throughout the region.

Because the seas were home to such abundant creatures, the sea floor accumulated quite a collection of remains as the creatures died. Over time—and here we refer to geological time, so we're talking about millions of years—all this material at the sea floor became compacted and hardened into limestone. The particular limestone formation in question became known as the Manitou Limestone for reasons which will become clear later.

Some 70 million years ago, the region underwent remarkable changes from shallow seas to the dry and mountainous region we know today. The Rocky Mountains were not formed primarily by volcanic activity as are many mountain ranges but by tectonic forces. The Pacific plate slid beneath the North American plate at a shallow angle, causing uplifts further inland from the plate margin than is typically seen. This resulted in in the American Rocky Mountains containing substantial amounts of granite, normally found much deeper in the crust. Because the lower plate slid shallowly beneath the upper plate rather than subducting deeply, it pushed the "basement" granite toward the surface, lifting it into a truly unique mountain range.

Then, around 7 million years ago, the aforementioned limestone deposits fell below the water table. When water (H_2O) mixes with carbon dioxide (CO_2), the dissolved mixture forms carbonic acid (H_2CO_3). Pools of this acidic mixture began to erode or "eat" holes in the relatively fragile limestone. As pools of rainwater formed in the resulting pockets, more carbonic acid formed, allowing the pools to eventually develop into long tunnels, forming the cave system.

1   More recent discoveries call this long-standing belief into question. It is possible that some vertebrates may have appeared as early as the Cambrian Period). However, it is much more certain that the late Ordovician saw the evolution of the first gnathostomes, which are the jawed vertebrates.

Some 4 million years ago, the water table dropped below the cave, which filled with oxygen. Mineral deposits such as stalactites and stalagmites began forming. These occur when water rich with minerals like calcium carbonate flows or drips through the cave, leaving some of the minerals behind. Over geologic time, these minerals accumulate to form the various cave structures. Stalactites form on the cave's ceiling as the water drips from above; stalagmites grow from the cave floor through the same process. Flowstone, curtain-like formations on the cave walls, form from deposits of calcite as water flows across the walls. Helictites, which look like clumps of worms or perhaps tree roots on the cave ceiling, form by a similar process, except their direction of growth shifts away from the vertical at some point in their formation for reasons which are not fully understood.

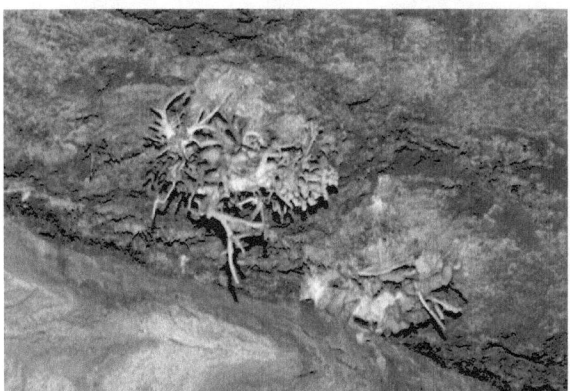

Figure 6.2. Cave of the Winds helictites. Photo: Bryan Bonner.

Though much of the known human history concerning the cave has to do with the Americans exploring the region during the Gold Rush era (and later), the cave is thought to have been known to both the Ute and the Apache. Of course, records regarding the extent to which these tribes explored the cave are sparse, but there are two legends worthy of note.

In Ute lore, there is a legend of a Chief and his daughter who hid in the archway entrance to the cave in an attempt to hide from attacking white men. It's unclear exactly how long they remained hidden there, but it must have been a long time, as they were later discovered by the daughter's lover. Both had died of starvation. Alas, little more is known about this legend. If true, all other details, including the Chief's identity, seem to have been lost to history.

The Apache also spent some time migrating through this region. In their lore, they speak of the Great Spirit of the Wind, responsible for whirlwinds, sand devils, tornadoes, and other windy sorts of disasters or phenomena. According to legend, they stumbled across a cavern that houses this Great Spirit of the Wind. That cave is none other than Cave of the Winds.

Figure 6.3. Cave of the Winds stalactites and stalagmites. Photo: Bryan Bonner.

In fact, the cave's name is taken from this Apache legend, and is also related to the cave's geology. Much like a whistle or musical instrument, the cave makes a moaning sound when the wind blows across the main natural entrance located on the cliff face of Williams Canyon.[2] It's quite possible—even likely—that this natural and haunting moaning sound is the original source of the Apache legend. It's not difficult to imagine that people predating our modern understanding of acoustics, upon hearing such a sound coming from a cave, would think its origin was some kind of spirit or monster.

The known history of the cave begins with very little fanfare in 1869 with a settler named Arthur B. Love who had established his homestead near the cave. He found and entered the main passage but never opened or explored the cave further. Exploration in earnest didn't begin for more than another decade.

In 1880, the Reverend Roselle T. Cross of the Congregational Church in Colorado Springs led a church-sponsored hike through the region. Among the participants were the brothers and schoolboys John and George Pickett. During the hike, they found a small shelter cave and noticed that their candles flickered while inside. They

---

2   Williams Canyon was given its name in 1870 by General William Palmer. It was originally called Manitou Canyon, but General Palmer changed its name in honor of his guide on the 1870 ride, William Truman Williams. Just to the northeast of Williams Canyon, one can find Queens Canyon, which was named for General Palmer's wife Queenie.

traced the source of the breeze to a small crevice. As curious young boys are wont to do, they immediately crawled through the crevice and discovered a large chamber on the other side. Reverend Cross immediately wrote up the discovery, publishing a description in his church newsletter, which was quickly picked up and reprinted by the *Colorado Springs Gazette* on July 2, 1880.

At least, that's the story of the cave's discovery you'll read on the website for the Cave of the Winds tourist attraction.[3] In reality, though that story isn't exactly false, the tale of the cave's discovery is a bit more convoluted, because different portions of the cave have been discovered by different people at different times, and some discoveries have been confused for neighboring (but disconnected) cave systems.

Parts of a cave system were discovered during limestone blasting in Williams Canyon as early as 1875 by Messrs. Case and Willcutt, who called this portion Mammoth Cave.[4] They opened the cave to public access before selling it to Tom Green in 1876. He maintained the tourist attraction business, charging fifty cents for admission, though he never bothered to develop it with such amenities as stairways or handrails. An early critic of Reverend Cross's claim to have discovered what became Cave of the Winds had actually confused Mammoth Cave (later renamed Hucacode Cave as a portmanteau of the names Hugh, Carl, Cora, and Delia) with the one the Pickett brothers had discovered. Reverend Cross was quick to defend his discovery against such detractors.

In 1880, following the Cross/Pickett discovery of one entrance to the cave, this portion, then called Cave of the Winds, was developed into a public attraction by Charles Cross and the Boynton Brothers. For their tour, they charted an entrance fee of $1 (just under $30 in 2022 dollars). However, apparently this was too steep a price, and the attraction closed later the same year.

Enter George Washington Snider, a stonecutter by trade who had recently settled in the Colorado Springs area from Ohio. In 1881, along with his friend Charles Rinehart, he discovered a portion of the system that came to be known as Canopy Hall. After excavating this remarkable cavern, Snider and Rinehart arranged to purchase the property and open a tourist attraction of their own, this time charging fifty cents for admission. Unfortunately, they were quick to announce the cave to the public and slow to develop any kind of security, with the result that careless or malicious individuals stripped the cavern of many of its remarkable geological formations such as stalactites and stalagmites. However, the commercial venture was a success, and established Cave of the Winds as one of the most popular tourist destinations in the region.

In 1885, Snider would also excavate the nearby Manitou Grand Caverns and open this cave system to the public as a show cave. At this early point, the public road to the cave wasn't yet open, so guests had to walk a 300-yard trail to visit the cave. But by

---

3   Cave of the Winds (n.d.). Park & Cave History. *Cave of the Winds*. <https://caveofthewinds.com/learning-center/cave-history/>

4   Not to be confused with Mammoth Cave National Park in Kentucky.

the early 1900s, Manitou Grand Caverns had been fitted with electric lights and made an exquisite show cave.

Other caves in the region (including the aforementioned Mammoth Cave, the Manitou Grand Caverns, as well as the Centipede Cave, later called Manitou Cave) have at varying times been open to the public as tourist destinations. However, only Cave of the Winds is still open for public tours to this day.

Snider's discoveries in and around Cave of the Winds complicate the history, however. In 1916, Snider challenged Reverend Cross's discovery of the Cave of the Winds, claiming he had personally been inside the cave as early as 1879 and found evidence that others had preceded him there. No evidence survives that can validate either Snider's or Cross's account of the true discovery of the cave.

Of course, either story is entirely plausible. Given that Mr. Love had found an entrance as early as 1869, even though he never explored it, it's not a stretch to think someone may very well have preceded Reverend Cross and the Pickett brothers into the cave. However, there's also reason to doubt Mr. Snider's claim to have personally visited in 1879. Given his zeal for exploring and commercializing caves in the region *after* 1880, one might expect that he would have been much noisier about his discovery if he'd actually found the cave *before* 1880.

Whatever the truth may be, there is no doubt that all of these individuals—Reverend Cross and the Picketts as well as George Snider—played an important role in the cave's history. There's also very little doubt that the Ute and the Apache preceded the lot of them into the cave, though the extent of their explorations is unknown.

Proceeding with some more firmly established history, we know that Snider operated his Manitou Grand Caverns tourist attraction in competition with the Cave of the Winds, which sparked a bitter rivalry between him and his onetime friend Mr. Rinehart. During this period, Snider's Grand Caverns business added several new attractions including the Hall of Crystal (now called Crystal Palace) and a monument dedicated to General Ulysses S. Grant, Abraham Lincoln, and General Robert E. Lee in 1885.

The rivalry between Rinehart and Snider culminated in litigation. Rinehart (and later his wife, Rose, on his behalf), began a series of lawsuits alleging the Manitou Grand Caverns should be jointly owned and managed by both Rinehart and Snider rather than just the latter. The litigation and subsequent appeals spanned the next decade, ultimately reaching the Colorado Supreme Court in 1894, though the high court declined to hear the case. The end result of the litigation was that, following Rinehart's death in 1895, his estate enjoyed a one-half stake in the Grand Caverns.

Snider, facing family debts and likely frustrated by the litigation, washed his hands of the entire affair. He left the business in the care of his brother Charles and left the area. The Grand Caverns continued its operation until eventually closing, most likely in or around 1906. While the Grand Caverns no longer exists as a tourist attraction, it was later connected to the Cave of the Winds by digging, and so it lives on as part of the larger Cave of the Winds system.

Either needing the money or having lost interest in the caves after the litigation, George Snider sold his interest in the caves in late 1916. He died on June 16, 1921 in Los Angeles, California. His body was returned to Colorado and buried in a grave in the family plot in Colorado Springs' Evergreen Cemetery.

Given that more than a century has passed since these events, there should be little wonder that there's plenty more history to discuss. However, most of the rest of the story consists of unrelated discoveries, projects, or vignettes rather than a connected narrative. Therefore, we present some of the noteworthy events from the cave's history, however briefly, in chronological order.

Another cave, now connected to the larger system, was discovered in 1892 by Perry Snider and Charles Austin. This cave, known as Middle Cave, is located between the Grand Caverns and the Cave of the Winds, and contains several large chambers adorned with spectacular geological formations. Some of the rooms in this cave, including the Valley of Dreams, Oriental Gardens, and the Adventure Room, are included in the current public tour of the Cave of the Winds.

Also in 1892, the caves saw their first wedding ceremony, as George Scholtz and Emma Thompson were married in what has become known as the Bridal Chamber (part of the Grand Caverns). The ceremony took place on May 4.

Amenities for tourists have been continually added over the years. On January 22, 1895, Temple Drive opened, allowing members of the public to drive directly to the new tunnel entrance, making the site more attractive to visitors.

As mentioned before, the Grand Caverns was outfitted with electric lights on October 11, 1904. Though this lighting was temporary and intended as part of a "woodmen of the world" convention, it marks the first time the cavern was electrically illuminated. This was not, however, the first cave in the world to receive electric lighting. Chifley Cave in Australia received that honor in 1880, when it was experimentally outfitted with electric lights.

In 1905, Charles Austin ordered the cave's first restoration project. W. O. Hooper, a cave employee, sprayed hydrochloric acid throughout the tour route to remove the black magnesium soot that had accumulated over the prior 20 years of touring and exploration.

Five new rooms, including the Crystal Palace and Finlay Hall, were added to the tours in February of 1906.

Also in 1906, the Grand Caverns saw its last tour groups. It did not reopen for visitors in the 1907 season as the management lacked funding to illuminate both caves and chose to focus only on the Cave of the Winds attraction.

On July 4, 1907, electric lights were permanently added to the caves. Thomas Edison, interested in the electric lighting system, visited the caves.

Another wedding in the Cave of the Winds, this time celebrating the marriage of

John Hague and Alta Davis, occurred on September 13, 1910.

In October of the same year, two unmarried women visited the cave. They asked the manager if they could add a small cigar box to one particular room in which unmarried women could place hair pins in the hope of finding husbands. Management agreed. According to unverified legend, the women returned the following year, both having been married in the interim.[5] The room is now known as the Old Maid's Kitchen.

In 1911, one of the cave's entrances was dynamited shut. There had been a problem with vandals damaging the property, so this was done in an attempt to control access to the cave and prevent further mischief.

Also in 1911, on July 10, an adjacent cave was sold to D. H. Rupp, J. F. Sanford, and R. D. Weir. They opened Centipede Cave to the public, charging fifty cents admission. This new attraction was next to the road leading to Cave of the Winds.

Attempts to prevent criminal mischief were not entirely successful. In October of 1912, burglars smashed open the entrance to Manitou Cave and stole all of the lamps along the tour route.

About a year later, on October 25, 1913, Charles Austin, manager of the Cave of the Winds property, purchased the 65 acre tract that contained Manitou Cave for $20,000 (equivalent to a little over $600,000 in 2022). Weir, Sanford, and Rupp sold the entirety of their interest in the region. This helped to consolidate ownership in the various competing cave properties (and cave tours) in the Williams Canyon region.

Further business development occurred in June of 1914 when local capitalists purchased Charles Austin's half of the interest in the Cave of the Winds property for $300,000. The remaining half was still at this time in the hands of George Snider and his family. Together, these parties organized a corporation to run the caves.

Another road providing easier access to Cave of the Winds, a serpentine drive from Manitou Springs, was completed on July 30, 1915, at a total cost of $8,000. From this point, the cave's history was fairly static for a period of several years.

Then, early in 1929, Ben Snider, nephew of none other than George Snider, jointly with cave employee Guy S. Boyd, discovered a new room in the cave while searching for a route between the Cave of the Winds and the Middle Cave. The new room was beyond the Rat Hole (now called Fat Man's Misery for reasons that are immediately clear to any larger man who attempts to cross this particular passage) at the end of the commercial section of the cave. Visitors were given the opportunity to suggest names for the new room, with the winner to receive a prize of $100. Over 25,000

5   Technically, the idea that placing a hair pin in a box in the Cave of the Winds could attract a husband qualifies as a paranormal claim and as such could have been placed in the following section of this chapter rather than in this historical discussion. However, we've decided that it's an untestable claim and because it's unrelated to most of the paranormal stories we'll discuss shortly, we present it here simply as part of the evolution of the cave's tourist program.

names were suggested, but the winning entry—The Temple of Silence—was submitted by none other than former owner Charles Austin and one Mrs. W. E. Greffenius of Fort Collins. They split the prize money. The room was opened to the public on May 22 of the same year.

Later in 1929, Ben Snider and Guy S. Boyd excavated a connection between Cave of the Winds, Middle Cave, and the Manitou Grand Caverns. Finally the once-competing caves were united into a single cave system operated under the name Cave of the Winds. The entrance to Middle Cave was sealed. Select groups were allowed to make the long journey from Cave of the Winds, through Middle Cave, to the historic Grand Caverns.

In 1935, the Valley of Dreams was opened to the public. A wall of this cavern bears an inscription attributed to George Jeffries, who is thought to have been an official from the Union Pacific Railroad. According to legend, he visited the cave prior to it becoming a commercial attraction and was so impressed he helped fund its opening as a tourist attraction. In return, he asked only for his inscription to be placed on the wall.

In the 1940s, during the Second World War, the Cave of the Winds was listed as an official air raid shelter by civil defense authorities.

On May 10, 1958, the Cave of the Winds Scenic Attractions Company granted a group of four corporations a 99-year lease on the cave operations at a cost of over $1 million.

The Serpentine Drive was opened to two way traffic in June of 1959. At the same time, lighting was added to Williams Canyon for nighttime viewing.

The lighting system on the tour route still used to the present day was added in 1974.

In 1980, the "Spelunker Tour" was reopened at the Grand Caverns for a price of $100.

On September 5, 1981, a time capsule was added to the Canopy Hall by Cave of the Winds owner Lou Carey and manager Grant Carey. Items in the capsule include a guide's shirt and badge, a $100 bill, a video cassette, pictures and brochures from the cave, a flashlight, a Bible, and two personal letters. The capsule is scheduled to be reopened 100 years after its installation. Since that will occur in 2081, there's a chance some of you reading this may still be around to witness the event.

Exciting news occurred on November 6, 1982: the discovery of the first new cave chamber since 1929. It was discovered by cavers from the Colorado Grotto of the National Speleological Society at the northern end of the Grand Concert Hall and came to be known as George Snider Hall.

Beginning on January 7, 1984, students and faculty from the Colorado School of Mines, the U. S. Air Force (whose Academy is located in nearby Colorado Springs), and the Southern Colorado Mountain Grotto (a chapter of the aforementioned National Speleological Society, a group of cave afficionados) underwent extensive explorations of the central portion of the cave system, discovering numerous previously uncharted passages. Ascending a sheer wall in the Shale's Belly cavern on January 14, the team discovered what is now called the Silent Splendor passage.

May of 1988 saw the opening of the new Adventure Room to members of the public. It was the first new room to be opened since 1881. However, it was eventually dropped from the tour route, becoming the first portion of the cave removed from the public tour since the lower level was closed in 1895.

Later that year, in November of 1988, a group of boys from the El Pueblo Boys Ranch in Pueblo, Colorado assisted experienced caver Gene Dover in excavating a short loop through what's called the Old Curiosity Shop to the north end of the Fat Man's Misery passage. In spring of the following year, the Old Curiosity Shop opened to members of the public.

Keen to keep offering new attractions to lure members of the public back to the cave, management added a "Laser Canyon" show beginning in May of 1991.

Similarly, in May of 1995, they added a "Lantern Tour" to their public tour offerings. It is still offered to this day. Instead of the fairly well-lit regular tour, this offering presents a dimmer view of the cave, coupled with the telling of ghost stories. It's a popular offering, and leads us directly into a discussion of the various ghost stories and paranormal claims surrounding the Cave of the Winds.

Paranormal Claims

Before we discuss the specific paranormal claims associated with Cave of the Winds, it's important to understand that the entire region is steeped in—and, indeed, named for—paranormal lore. Manitou Springs, Colorado takes its name from the legend of the manitou.

In the Native American theology of the Algonquian peoples, the manitou is the omnipresent spirit or fundamental life force. Though horror fans are probably most familiar with the concept of the manitou through Graham Masterton's 1976 novel *The Manitou* or its 1978 film adaptation of the same title[6], the original concept of the manitou is not necessarily of a malevolent spirit, but merely of a powerful spirit.

Similarly, as we already mentioned, even the name of Cave of the Winds, according to some legends, comes from Apache lore that the cave system was home to the Great Spirit of the Wind, likely due to the haunting moaning sounds emanating from the caves.

However, because we now understand how the moaning sounds may be produced by the cave's acoustics and because we're in the business of looking into specific and testable paranormal claims rather than religious lore (though the two may sometimes

6   Or, indeed, Peter Straub's classic 1979 novel *Ghost Story*, in which the concept of a manitou plays a large role. However, the same cannot be said of the 1981 film adaptation which presents more of a traditional ghost story than Straub's novel.

overlap), we took a greater interest in the specific paranormal phenomena people have reported witnessing within the cave's chambers.

Numerous tour guides and some of their guests have reported seeing ghostly apparitions within the cave. On some occasions, while guiding a tour group through the tunnels, they've reported seeing extra guides (complete with the standard uniform) who weren't meant to be there. On other occasions, extra guests were seen on the tour even after all the guests and guides had been accounted for.

Our contact at the Cave of the Winds told us of a journey in which a group was on a caving trip to the Grand Cavern. As they walked, climbed, crawled, and shuffled through the cave's passages, they began to hear sounds behind them of footsteps and voices. Believing another caving group was slowly approaching from behind, they reversed course temporarily to go introduce themselves. But no one else was in the cave. Despite extensive searching, they never found another group.

The same individual told us of another occasion in which he was walking one of the cave's paths. Up ahead, he saw another person standing on the path, backlit by a lantern from the other side of the path. Thus presented in silhouette, the individual wasn't easily recognizable, and our contact couldn't make out any identifying features. But while he walked toward the individual, the person "just vanished."

Yet another story relayed to us by our contact involved an individual who claimed to be "sensitive" to spirit or paranormal phenomena. While visiting the cave, she attempted communication with the resident spirits and started relaying detailed information about the cave, including the locations of several "lost tunnels" as well as names of the original founders of the cave. Reportedly, she had no prior knowledge of the cave or its history.

For safety, there are several emergency call boxes strategically placed throughout the caves, to allow for communication between the tour groups and those outside the caves back in the office. Over the years, there have been numerous reports from people who've heard these phones ringing in the caves even when nobody was on a tour. Others, trying a bit of paranormal investigation on their own, have left some of these phones off the hook to listen to the sounds of the empty caves only to hear the sounds of people walking through even when the caves were known to be empty.

Even the attraction's gift shop has some haunting stories. Apparently their phones will ring from time to time even when they're not plugged in. Additionally, some people have reported seeing "shadow people" at the gift shop counter.

Cave staff and visitors have reported seeing spirits of Native Americans wandering through the caves.

The one spirit most precisely identified in the lore is that of the man who discovered the caves. Reports are not clear as to whether this refers to Mr. Love, Reverend Cross, Mr. Snider, one of the Pickett brothers, or one of the other people who have discovered portions of these caves. Regardless, this individual is sometimes seen wandering either near the original entrance or around the monuments. Perhaps the spirit's reported presence around the monuments might suggest this individual is meant to be George Snider. Indeed, though specifics are generally sparse, it's been widely reported that the Cave of the Winds is haunted by the spirits of George Snider (though his name is sometimes misreported in these stories as George Pickett) and his wife Nellie.

Mr. Snider himself features in a couple ghost stories from the cave's history. According to the legend, during the late 1880s when Mr. Snider was busy exploring the caves and building them up as a tourist destination, he would often host private parties within the cave system. Because the caves hadn't yet been fully developed into an easily accessible attraction, Mr. Snider took to paying local boys five cents per hour to carry lanterns and bags into the caves for his guests. On one occasion, after cleaning up from a party, a distraught mother approached Mr. Snider and told him that her sons had never returned. As the story goes, Mr. Snider didn't entirely believe her because he'd already checked that all his guests departed safely. Nevertheless, he undertook another search of the caves and found them empty. No sign of the missing boys was ever found, in the caves or anywhere else. But according to the story, one can occasionally still hear the ghostly sounds of children laughing and playing in the cave's tunnels.

Another story suggests that Mr. Snider had a bit of the prankster spirit about him and enjoyed giving his guests a nice good-natured fright when they explored the caves. According to this legend, one of Mr. Snider's brothers unearthed three human mummies (presumably Native American) during a nearby mining operation, and Mr. Snider contrived to purchase one of them. He hid the mummy somewhere in the cave and used it to give his guests a good scare. One day, though, he received word that some of his guests already knew about the mummy in advance, so he arranged to move it to a different location than the one in which they were expecting to find it. However, when he went to go retrieve the mummy and move it to its new location, it was gone. As Mr. Snider was the only one with keys to the cave's entrance, he couldn't think of anyone who might have been able to pull off the heist, yet the mummy was never found.

A story entitled "Whispering Walls" was included as part of Season 4, Episode 3 of the popular paranormal television program *My Ghost Story* in 2012. Though this was after our own investigation, it's worthy of note that this story does align with some of the established ghost stories. The show's subjects reported a successful EVP session with a spirit they believe to be that of Nellie Snider who said "I'm coming out of the wall," shortly before they captured a blurry photograph that appeared to depict a shape near the cave wall. They later recorded another EVP which said, "I'm fine; it's okay." Deeper in the cave, they reported interacting with a spirit that felt somehow more malevolent near a steep ladder where they'd been told someone had died from a fall many decades prior. Because these were reported after our investigation and we weren't present for any of the experiences or recordings, we're not including them in our investigation report, but choose to mention the stories here merely in the interest of documenting the variety of stories that have been reported over the years.

Finally, we'd be remiss if we didn't mention perhaps the most famous of all paranormal claims ever made about Cave of the Winds. According to an episode of the television program *South Park*, Cave of the Winds is home to a cryptid known as ManBearPig (half man, half bear, half pig). It may just be a fictional story from a comedic television program, but we assume that any mention of Cave of the Winds (at least among certain circles) is sure to provoke mention of the beast, so we figured we'd get ahead of matters and get that out of the way. As far as we know, no one (except maybe Al Gore) actually believes in ManBearPig.[7]

Of course, many of these stories are the kinds of one-time events that we both love and hate. We love them because they're excellent ghost stories (and we wouldn't be in this business if we weren't complete suckers for a good ghost story). But we also hate them because their nature as one-time events makes them difficult to impossible to examine forensically or scientifically, and we're left with just having to take the stories for what they are. Nevertheless, Cave of the Winds was home to so many ghost stories, reported by so many different people, we were keen to undertake an investigation. It is to that investigation that we now turn our attention.

Our Investigation

Every investigation presents its challenges. Investigating in an active hotel (or worse, an active resort and casino—and yes, we've done it) presents obvious challenges when it comes to noise or interference from the public. Trying to work in a cemetery presents challenges related to being outdoors and exposed to the elements. Private residences are often challenging for a wide variety of reasons to do with either the conditions of the property or the simple lack of accessible historical information about the property. However, nothing quite prepared us for the unique challenges involved in conducting a paranormal investigation inside a large cave system.

Indeed, our ability to work in this unique setting, though it presented some frustrating obstacles, also made this one of our favorite investigations simply because it was so different from all the rest.

One of the biggest challenges we had to think about during the planning stages of our investigation was the sheer size of the property. On a typical investigation, we try not to bring too many people with us. Large numbers tend to breed confusion and the potential for contaminated data. It's essential for us to be able to keep track of where everyone is and what they're doing at all times so if, for instance, we record anomalous audio, we're able to easily determine whether the source might have been one of our own people. In the case of Cave of the Winds, though, the site was simply too big for our usual team to cover all the relevant ground, so we invited another

7   We're super cereal.

(now defunct, as far as we've been able to learn) paranormal group known as Spirit Paranormal Investigations, or SpiritPI. We'd encountered several of their people over the years and knew that, though their methods were not identical to our own, they were a dedicated group of people who would understand that we were there to work, not merely to play (not that we're averse to having some fun along the way as long as we're careful to prioritize the science over the shenanigans).

The next challenge was closely related to the first one. Cave of the Winds is big. And though it does have electric lighting throughout much of the cave, it doesn't have power outlets throughout nor even electric lighting in some of the regions of interest to us. Worse, some of the passages are quite small, and we carry with us a *lot* of equipment. Worse still, the areas of interest to us were not located right near the cave entrance. No, we had to move deep into the cave. Moving our equipment in and getting set up, a process that usually takes us mere minutes up to perhaps an hour (depending on the size of a location), became one of the most labor-intensive projects not only during this particular investigation but during our decades of work.

Recall from earlier that there's a passage in the caves called Fat Man's Misery. Its name is no accident. Though this particular passage was not part of our route, its name is indicative of the kind of conditions with which we had to contend. At one point, we had to pass through an opening 42 inches wide and 50 inches tall. While this is easy enough to pass through as long as one's back and knees are up to a stooping walk, it's not so easy to pass through while carrying large cases filled with heavy equipment. Indeed, several cases had to be left behind (and their contents carried in separately) because they simply wouldn't fit.

Once the investigation started, we had to occasionally find our way back through these passages whenever we had to use the restroom. Not the most comfortable of journeys, but also a remarkable experience. Traveling alone on these "missions," some of our people took a moment to turn out the lights and enjoy a few minutes of the kind of solitude, darkness, and silence many people never get to experience. That alone would have made all the trouble worthwhile. Incidentally, though the Cave of the Winds is not home to an enormous bat population like some caves are, a small bat did follow one of our people back into the cave and briefly join us on our investigation.

After getting all the equipment inside (and perhaps inventing one or two new curse words in the process), we faced another challenge. The vast majority of our equipment requires electricity. True, we have plenty of handheld gizmos that operate on battery power, but that's simply not enough for a proper investigation. We needed to find a way to power our tools. We proceeded to run more about a sixth of a mile of extension cord into the cave interior.

With those challenges out of the way, we were finally able to begin setting up our monitoring equipment in much the same way that we normally would. We set up four infrared-capable cameras[8] connected to a DVR and monitor along with a still camera, several digital recorders, and established locations for EMF readings throughout the Monument area of the cave.

The area our investigation covered was similar to the area covered by the "haunted" Lantern Tour still offered by the cave and focused on the portion of the system that was once the Manitou Grand Caverns.

One of the areas we wanted to monitor was known as Lovers Lane. In that section, we placed one microphone connected to a computer to record audio. For video, we used one of the infrared-capable security cameras connected to the DVR. We also established two fixed locations in this region for repeated EMF readings throughout the investigation.

In the Concert Hall, we placed two of the infrared capable video cameras as well as two microphones connected directly to a monitor. This region also featured two locations for repeated EMF readings.

The investigation lasted about seven hours. Throughout the evening, temperatures remained consistent at approximately 54 degrees Fahrenheit, with nearly 100% humidity. Readings on the EMF meters were underwhelming. We found no anomalies. Indeed, with the exception of our own equipment, we found no electromagnetic fields at all.[9]

We were also disappointed that we weren't able to pick up anything interesting on any of our video or audio recording equipment.

However, the trip wasn't entirely a bust. Several of our team members reported hearing something they described as sounding like a violin in the Concert Hall room. Specifically, they said it sounded like it was coming from behind the walls of the cave.

Several other members reported seeing "something" moving at the back of Lover's Lane and hearing "shuffling" sounds in or around the same area.

---

8   In some paranormal circles, people seem to be of the opinion that infrared cameras are more likely to detect ghostly activity than those operating with visible light. This has never been demonstrated to our satisfaction. However, when one is setting up video monitoring in a low-light setting such as a cave, it's necessary to have infrared capability simply to ensure the cameras are still able to see in the semi-dark conditions.

9   Importantly, there are two kinds of EMF. Manmade objects such as computers, power lines, appliances, etc., produce alternating current (AC) electromagnetic fields. Most commercially available EMF meters look for this kind of EMF. However, natural direct current (DC) magnetic fields do exist and meters are available that detect them. However, because these devices are designed to detect small fluctuations in the Earth's natural magnetic field, they are highly sensitive and those not professionally trained in their use are likely to detect themselves or their teammates rather than whatever they're looking for. We've even seen them detect an individual walking down a hallway on the other side of a building. In this investigation, we were using AC EMF meters, which is why we didn't find much of a field deep in the cave.

We were never able to formally document any of these phenomena on our recording equipment nor determine their possible sources. Could they be ghosts? Strange acoustic phenomena because of the unfamiliar environment? Though we can hypothesize, we weren't able to conclusively solve the mystery.

One of the things that's worthy of consideration is that caves are naturally disconcerting places. If you turn off whatever lights you've brought with you, you'll find yourself in the kind of darkness most people never experience. Especially for those of us who live in urban or suburban areas, light is ubiquitous.[10] Even when we turn out the lights to go to sleep at night, it's still possible to see *something*, albeit not as clearly as it would be in daylight. To a lesser extent, the same is true in rural areas. Moonlight, starlight, and other light sources, however faint or distant, still provide some illumination. In the cave, if you turn the lights out, no matter how long you let your eyes adjust, you'll never be able to see anything. You can't even see your own hand an inch in front of your face.

Figure 6.4. Members of the team check up on some monitoring equipment during their rounds.
Photo: Bryan Bonner.

Psychologists have studied sensory deprivation over the years and determined that this kind of darkness plays all kinds of tricks on the mind. Just a few minutes of total sensory deprivation is sufficient for many people to hallucinate. Several hours of sensory deprivation will make almost everyone hallucinate.

Granted, though we turned the lights out once or twice just to experience the kind of darkness we don't get to see every day, we did most of our work in the light.

---

10   To the eternal frustration of astronomers and other stargazers. If you've never done so before, compare the night sky in an urban or suburban area to that in a rural area. It will be an extraordinarily educational experience.

We were investigating, not conducting a sensory deprivation experiment.[11] However, the environment was still quite disorienting. Even with the lights on, the cave was relatively dim. Sound carries differently in such an environment. It's impossible to orient oneself with respect to direction.

Though we've trained ourselves to be quite adept at separating figments of our imagination from real observations, it's entirely possible that some of the experiences our members had may have been due to the bizarre environment, in one of two ways. First, it's possible they actually did see and hear things, but these were the results of natural phenomena behaving in unexpected ways due to the cave's unfamiliar conditions. For instance, the sounds of violins could have been natural sounds or sounds of our own people simply carrying strangely through the cave, producing a sound that only *seemed* like music. Second, it's also possible the conditions could have caused minor and brief visual and auditory hallucinations.

Importantly, we are not asserting that these are the explanations. We're merely mentioning them as hypotheticals in the absence of further evidence. Some might assume our people heard or saw ghosts. Others might assume it was some trick of the cave's geology. One or two of our people even jokingly suggested it was ManBearPig. Unfortunately, since we found no further evidence, we have to leave that as an unknown in our case files.

But though we may not have solved the mystery, it did capture our attention, and we knew there was more work we could do, so we arranged for a second visit to the cave.

For the second visit, we decided to travel lighter. Instead of bringing the entire set-up, we elected to pack only laptop computers, small video cameras, and handheld devices. Naturally, that represents something of a tradeoff. Normally, we like to overequip ourselves. The idea is that, since the paranormal is an ill-defined field and no one actually knows exactly what they're looking for, we want to measure, record, or document as much as possible in as many different ways as possible to give ourselves the best chance of finding something interesting. However, given the physical constraints of investigation in the cave, we decided it would be better to lose some of the equipment and gain all the time that would otherwise have been spent loading in and out.

On this occasion, we arrived at about 7:00 p.m., and it only took one trip to the back of the cave to carry in all of the equipment. We set up video cameras in Lovers' Lane, the Concert Hall, and the Monument Room as well as microphones in Lovers' Lane and the Concert Hall. We established EMF metering stations at specific points in all three locations. In all, we used 12 cameras, 2 thermometers, 3 EMF meters, and a team of six people on this second investigation, with our base camp established

---

11 Some members of Rocky Mountain Paranormal have conducted personal sensory deprivation experiments to help ourselves understand the psychological effects of such alien environments, but those were conducted under controlled conditions and were not directly related to this investigation.

between the Concert Hall and Lovers' Lane.

After an initial uneventful quiet phase, we decided to send two members to the back of Lovers Lane. These individuals spent about twenty minutes there, and reported back to the team that they'd heard a sound they described as akin to a motor running somewhere in the cave. When a third person returned with them to the location, the sound was no longer heard.

This is another one of those frustrating occurrences that happened so briefly we were unable to figure out exactly what the source of the sound was, whether normal or paranormal. At the very least, we can confirm that no one was operating motorized equipment in the caves during our investigation, but beyond that, we can only speculate.

A bit later, one of the team members reported seeing "lights at the back of the tunnel." After a time, another of our crew reported seeing the same phenomenon. There were no tour groups during our investigation and no one else in the caves who might have had flashlights, so this was a rather exciting discovery for a while.

However, rather than immediately jumping to the paranormal conclusion, we kept searching for possible explanations and believe we have plausibly (if not conclusively) solved this mystery. There's a well-documented psychological phenomenon known as "prisoner's cinema," which got its name after it was initially described in prisoners who'd been confined to dark cells for long periods of time. What psychologists have figured out is that when people (whether voluntarily or involuntarily) spend long periods of time in the dark, they see a "cinema" consisting of a sort of "light show" of various colors that appear out of the darkness.

The lights in this so-called light show are often difficult for individuals to describe, though they sometimes take the form of human faces or other figures. This latter point is likely related to another well-documented psychological phenomenon known as pareidolia, in which people see meaningful patterns (particularly human faces) in random or meaningless visual noise.

Phosphenes are "lights" that are perceived even when no actual light enters the retina. They can occur under a wide variety of circumstances. The prisoner's cinema is thought to be a combination of these phenomena with the other psychological effects of prolonged lack of light. It has been reported by prisoners, truck drivers, pilots, astronauts, and practitioners of intense meditation.

Indeed, some people can even produce these kinds of hallucinations on a voluntary basis without too much trouble. As a child, one of this book's authors used to enjoy keeping his eyes open for a long time in a darkened room before going to sleep and would often see illusory patterns of green lights morph in front of his eyes.

While we can't conclusively prove that the prisoner's cinema effect is the explanation for the lights our team members witnessed, it seems like the most plausible explanation.

Most of the rest of the investigation proceeded uneventfully and consisted of

quiet periods punctuated by hourly EMF sweeps. During these breaks, different vol-
unteers would relocate to different regions of caves under surveillance to see if they
could find anything of interest. For the most part, they did not.

One team member did see one last thing worthy of note, though. Upon returning from a break, this individual reported seeing a man standing in the small tunnel
to the left of the entrance to the Monument Room. When he went to investigate, he
couldn't find anyone. All of our team members' locations were accounted for. Was
this the spirit of some long-deceased cave enthusiast? A hallucination? A trick of the
dim lighting? We've never been able to find out for certain.

At the end of the investigation, we packed up our things, grateful for the remarkable experience of spending some time in these wonderful geological marvels but little
closer to proving or disproving any of the paranormal claims.

South Park fans will also be disappointed to learn that we found no evidence of
ManBearPig.

References and Further Reading

Cave of the Winds (n.d.). Park & Cave History. *Cave of the Winds.* <https://caveofthewinds.
        com/learning-center/cave-history/>
Lamb. K. (2016). *Ghosthunting Colorado.* Covington, KY: Clerisy Press.

7
A Haunted Former Orphanage: The McClelland School

The McClelland School, located at 415 East Abriendo Avenue in Pueblo, Colorado, is a small private pre-kindergarten through eighth grade institution offering the benefits of a private education and small class sizes (the entire school serves fewer than 100 students) to children in its region. But it wasn't always a house of learning. The building was once an orphanage. While there are no claims of abuse or mistreatment of the children, it was not always a happy place. Understandably, it has a haunted reputation to this day.

Figure 7.1. The McClelland School. Photo: Bryan Bonner.

The History

The story of the McClelland School begins long before anyone was thinking of opening a school. By the late 1870s, the city of Pueblo, about 100 miles south of Denver, was becoming one of Colorado's major economic hubs. As the town grew, a group of concerned citizens known as the Protestant Orphanage Committee met on January 10, 1905 to discuss the need for a local orphanage. Almost immediately, Andrew J. McClelland entered the picture.

Mr. McClelland was born in Virginia in 1850 and had been a teacher for a number of years before heading toward the promise of riches offered in early Colorado. After establishing a grain business in Georgetown, he sold his share in the company in 1881 and eventually settled in Pueblo in 1882. He would go on to become one of the most prominent figures in the town's early history.

He maintained flour and grain businesses but also took a keen interest in real estate, even naming the Columbia Heights subdivision after his wife, Columbia Jane McClelland. His influence and philanthropy did a lot for the growing city. In 1887, he brought the Missouri Pacific Railroad to Pueblo. In 1891, he founded the town's first public library, which bears his name. In 1904, following a trip around the world, he returned with a collection of historic and cultural treasures, most of which are now housed at the Rosemount Victorian House.[1] He also published *The Pueblo Leader*, a local newspaper.

And after the group of Pueblo citizens decided the city needed an orphanage, he donated a building to the cause: a three-story structure that had formerly been a college affiliated with the Southern Methodist Church.

Before we return to the story of the building itself, it's worth pointing out that apparently no good deed goes unpunished. McClelland, one of the region's most prominent businessmen and philanthropists, was an obvious choice to become involved in local politics. In this chapter in his life, he found an enemy in Howard Zink, a local politician who blamed McClelland for the dirty business taking place in the town's red light district—at that time, Pueblo was home to numerous saloons, brothels, gambling halls, and the like, and efforts at reform had largely failed.

The public opinion against McClelland reached the point that Columbia McClelland had to appear at a city council meeting to speak on behalf of her husband.

"If we are such undesirable citizens," she said, "give us back our money and we will leave."

Apparently that silenced the rabble, at least to an extent.

1   McClelland's trip to Egypt resulted in the acquisition of two mummies. These are now on permanent loan to the Denver Museum of Nature and Science, where they are prominently displayed as part of the Egyptian Mummies gallery and are still being studied today.

However, Zink's attacks continued. At one point, he attempted unsuccessfully to have McClelland removed from the library board. At another point, McClelland's political enemies attempted to have him prosecuted for allegedly sending obscene material through the mail.

Eventually, enough was enough. McClelland left the area and eventually settled in Pasadena, California. Columbia remained in Pueblo. It's likely that the couple had some marital troubles that may have necessitated a separation. However, it's also worthy of note that the separation appeared to be reasonably amicable, as Columbia continued to manage the couple's business affairs in Pueblo.

Andrew McClelland died in Pasadena in March of 1936. He was 86 years old. His body is buried in Pueblo's Roselawn Cemetery. Columbia Jane McClelland died in 1939 and is buried next to her husband.

The McClellands had no children of their own, which may be part of the reason they were so interested in supporting the other children in Pueblo. Whatever the reason, the orphanage would not have been established without their money, support, and influence. The McClelland Orphanage, bearing the name of its benefactors, opened its doors on April 25, 1906, and was designed to house as many as 65 children at a time.

Though modern readers may have a particular image of what an orphanage would have been like at the turn of the previous century, children were placed at the orphanage for numerous reasons—including the deaths of the parents but also including separations of the parents—but not all of the placements were permanent. In fact, most children remained only a month or two before being reclaimed by the parent or parents, presumably after life had become more settled. Infants placed at the orphanage awaiting adoption typically didn't have long to wait. However, the orphanage was also home to some children (though a minority) who were never placed with an outside family.

In 1927, the institution changed its name from The McClelland Orphanage to The McClelland Children's Home. For the sake of simplicity, we will continue to use the word "orphanage" in this chapter.

Interestingly, the building that stands today was not the original structure. Plans to replace the building with a new one began circulating in 1932. However, this was in the midst of the Great Depression, and funds were dangerously low. In response, most of the orphanage's holdings were liquidated. Combined with a donation of $25,000 from the Pueblo Rotary Club, this was just enough to fund the renovation. Construction on the new building began in 1934 and was finished the following year.

Life in the orphanage was not necessarily easy. However, those of us who grew up reading Charles Dickens novels may not have the right idea about conditions at the McClelland Orphanage. In fact, the institution was widely praised for being among the best-run and most humane of such institutions in its day. The people in charge did a remarkable job, and most of the orphanage's wards were, by all accounts, at least content with their lives, if not exactly happy.

But there are always outliers. A story (admittedly unconfirmed) has it that one of the wards repeatedly tried to escape through the upper floor window in the hopes of finding her way (presumably by train) to Trinidad to reunite with her mother. Beyond this, there were the occasional late-night runaways. But by and large, the children were content.

They were also quite close. This builds bonds, but it also spreads illness. Despite the best efforts—and we mean "best efforts" sincerely in this case—of the staff and administration of the orphanage, former wards of the home recalled years later that whenever one child got sick, the illness would spread rapidly through the entire population.[2]

Medicine at the time when the orphanage was in operation was not what it is today. Unfortunately, this means the building has certainly seen more than its share of death. Even more unfortunately, these deaths were those of children.

Conditions at the orphanage remained fairly consistent for most of the time. By the late 1960s, the need for an orphanage in Pueblo had begun to decline. In place of orphanages, society moved toward something more akin to our current model of child welfare programs. The trustees of the orphanage began to change their focus and began shifting their services away from those of an orphanage and toward those of an educational center. Specifically, they focused on the diagnosis and evaluation of the learning process in children.

In 1973, the institution became The McClelland Center for Child Study. This coincided with a complete renovation and redecoration of the building, both inside and out. Funded by a grant to operate a preschool program to study early detection of potential learning disabilities, the program began its first full year of operation in 1975. This grant lasted until 1977. By 1979, this program had grown considerably and served a core group of twenty students in kindergarten through third grade. Plans were made to expand the program to include fourth and fifth grade educational services.

Matters remained fairly consistent for the next fifteen years, marked by only gradual growth. The school continued to offer pre-kindergarten through fifth grade services. As enrollments grew, the school gradually added additional classrooms and eventually a gymnasium.

Having become more of a traditional private school than a research facility, the institution changed its name from The McClelland Center for Child Study to The McClelland School in 1994. Middle school services were added in 1995, and the school now offers classes at the pre-kindergarten through eighth grade level.

From 1996 to 1997, following a capital fund campaign to raise the necessary money, the school renovated the old orphanage carriage house in order to create a separate

2   Purfield, T. (2005). A Child's Place. *The Pueblo Chieftain.* <https://www.chieftain.com/story/news/2005/07/11/a-child-s-place/8470605007/>

facility for the middle school students. This building is now the primary classroom area for students of middle school age.

At this point, the school serves just under 100 students at any time. In addition to the standard academic curriculum offered at any school, their curriculum also includes Spanish language, music, art, and physical education for students at all grade levels.

Paranormal Claims

Schools can be inherently creepy places. During the day, they're centers of learning, full of life and the sounds of youth. At night, these large empty structures just seem…wrong, somehow. Perhaps it's because we're so conditioned to expect schools to be filled with light and action and the sounds of children that they seem alien when empty, dark, and silent. It's little wonder that many schools have developed a haunted reputation over the years.

In the case of The McClelland School, formerly an orphanage, it seems like a no-brainer that there'd be some ghost stories. And indeed there are.

As is the case with most (but not all) locations, the ghost stories are collections of disjointed anecdotes rather than coherent narratives. Students and staff at the school have reported a variety of anomalous experiences. Many of these stories follow a similar format and structure to the ghost stories at other schools or allegedly haunted locations.

Several students have reported over the years that they've seen children playing on the school grounds even when school was in session and no students were supposed to be out playing. One's inner skeptic may suspect truancy is a more likely explanation than ghostly activity, but the story has been reported specifically as a ghost story often enough that it merits consideration.

Some teachers have reported hearing the sounds of entire classes of children entering the building. Upon investigation, they found no one. These claims got us a little more excited because it's substantially harder to explain how one might hear a large group of children *inside* the building than one or two outside of it.

The school's music room is location of particular interest. Multiple teachers have reported odd knocking sounds coming from the ceiling above this room.

In addition to the music *room*, there's also a ghost story of a music *box*, which can reportedly been heard playing on the school's third floor. There is no music box there.

One local resident claimed that as he walked past the school one evening, he noticed the lights were on uncharacteristically late at night. He approached the school and peeked in the window to verify everything was alright. Inside, he saw a man sitting in a rocking chair by the fireplace in the building's lower level.

Even school staff have reported some of these visual manifestations. One teacher reported seeing a pair of legs (not attached to the remainder of a body) "sitting" on the third story stairs.

Indeed, the third floor seems to be a hotbed of claimed paranormal activity. On this floor, there's a particular room assigned at the time of our investigation to a teacher named Ms. G——. She told stories of hearing children when none were present, smelling cigarette smoke, and hearing a screen door close, even though there is no such door in her room.

The school's caretaker also added some ghost stories, claiming to have heard the sounds of children crying after hours when no children were present, as well as hearing the latches on the doors rattling even when he was alone in the building.

And that's the most common sort of claim made of paranormal activity at The McClelland School—sounds of children even when no children are present. These occurrences have been reported by multiple people and at locations throughout the school and its grounds.

Stories like these didn't give us the most information to go on, but we're accustomed to beginning our investigations with limited information and elected to see what we could discover.

Our Investigation

After doing a bit of background research to learn about the location's history, one of the first things we wanted to know was whether we could verify any deaths at The McClelland School (either as a school or, more likely, when it was still an orphanage). Paranormal studies are ill-defined, so there's no definitive link between death at a location and the presence of (alleged) ghosts, but there does seem at least to be a strong correlation between the two. Lore suggests that places where many people died (or even a few people, if they died horribly or prematurely) are more likely to be haunted.

In this case, we were fortunate enough to be able to find records of some of the deaths that occurred in the building. Unfortunately, all of the deaths (at least as far as we were able to discover) were of children.

While conducting this research, we also found a lock of hair belonging to "Tony" in the records from over a century ago.

The following represents all of the known deaths that occurred at The McClelland Orphanage:

- Arlui Boling – Born January 5, 1895; Died November 30, 1907,
- Mildred McClelland – Found May 15, 1908; Died July 3, 1908,
- Amie Marie Anderson – Born July 29, 1908; Died January 3, 1909,
- Edward Smith – Born September 30, 1924; Died January 26, 1925 (from "intestinal flu"),
- Mariah Johnson – Born July 12, 1913; Died June 20, 1925 (from appendicitis),
- Morris Jackson – Born May 15, 1925; Died September 20, 1925,

- Milfred Sharp – Born May 29, 1931; Died December 25, 1931 (from "thymus gland trouble," and discovered dead in the morning),
- Ruby Santrelli – Born July 3, 1922; Died September 3, 1932, and
- Paul Floyd – Born January 29, 1928; Died August 8, 1941.

It's important to note that, though these were the only deaths in the building we were able to confirm and to which we were able to attach names, they do not represent a complete list of deaths in the orphanage. Likely there are many others, but we have no way of knowing precisely how many.

For our on-site investigation, we established a base camp on the second floor stairwell and, as is our way, placed monitoring equipment throughout the facility. Wireless cameras were placed in the second floor hallway, the second floor classroom, and two in the basement. One microphone was placed in the basement.

As is also our way, we began by taking baseline EMF and temperature readings and repeating these measurements periodically throughout the evening.

At about 7:50 p.m., we discovered the lens cover on a camcorder placed in the basement had been physically closed. This particular camera has a lens cover that has to be manually closed. That is, even if the device loses power, the lens cover remains open. We had previously verified that the cameras were working, so someone (or perhaps some*thing*) had to do this manually. We never caught the culprit.

Two of our members reported hearing what they described as sounding like a little girl singing or humming at around 10:30 p.m. Nothing was heard on the audio recording. That two of our team members heard this sound independently suggests that the sound was real and not just our minds playing tricks on us[3], though we never figured out its source.

More excitement developed around 11:00 p.m. when we collected abnormally high EMF readings on the third floor stairwell landing. These readings fluctuated and appeared to change location. This mystery, however, we were able to solve, and we were able to attribute the unusual readings to a corridor light behind an adjoining wall.

At precisely 12:49 p.m., all members of our team heard a loud banging or slamming sound. Initially, no one could identify its source, so we divided into teams to investigate. This represented a deviation from our usual "sit down and shut up" protocol

---

3   This is something of which we have to constantly be aware. When you go to creepy old places in the middle of the night and try to investigate, sometimes sleep-deprived, after spending hours listening to people tell their ghost stories, it's quite easy to "psych yourself out" or imagine seeing or hearing things that might not really be there. Even with all of our training and experience, we are still vulnerable to the foibles of human psychology, so we're always keen to seek independent verification of any phenomenon.

to avoid contaminating our own data, but when anomalous events occur, we have no problem temporarily abandoning standard procedures to look into the new occurrence.

Eventually, we found what we believe to be the culprit, but this answer simply presents new questions. A book—volume "I" of a set of encyclopedias—which had been on a bookshelf in a second floor classroom at the beginning of the evening was resting on the floor. Several other books were in different places on the shelf than they had been previously.

Figure 7.2. This book fell off the bookcase, propelled by an unknown force. Photo: Bryan Bonner.

Here is where paranormal investigation is simultaneously the best and worst job in the world. In this case, we've just witnessed something that many people would attribute to ghostly activity. While we're cautious not to go quite that far and choose instead to call it a mystery, it's the kind of thing that gets us really excited. Many investigations come and go with no unusual happenings at all, but we live for these kinds of cases in which things actually do occur. And yet, it's also the most frustrating thing in the world, because we simply can't say definitively what happened. By all means, we're careful to control circumstances on our investigations to the greatest extent possible, so we don't think there was any opportunity for some prankster to pull a joke on us, but at the same time, all we really have as evidence is a book out of place, and that's just insufficient to conclusively prove anything paranormal.

That doesn't mean we didn't try. After carefully photographing the scene, we replaced the book on the shelf and began experimenting with the amount of force that would be required to remove the book from the bookcase. We did notice that the shelf was wider than the book's height, suggesting that if the book had simply fallen over, it would have remained on the shelf. For it to appear on the floor instead means it likely required an applied force (as opposed to gravitational force alone) to get where it ended up. The case of the fallen book remains an unsolved mystery.

Things continued to occur as the evening went on. At 1:05 a.m., just a few minutes after the incident of the fallen book, another loud bang echoed from a lower floor. Our teams immediately sprang into action to determine the source of the sound but were unsuccessful. Nothing was found out of place anywhere in the school, and we were never able to determine the source of the noise.

At 1:07 a.m., two individuals (one member of our team and one staff member of the school) reported they saw a woman walking into a third floor room on one of the monitors. Immediately, we separated the two witnesses.

This is an important step, which we perform not because we doubt their integrity, but because we want to ensure that neither one can—however unconsciously or unwittingly—influence the story told by the other. When two people tell the same story one after the other, the evidence is far less convincing than if two people tell the same story independently, without knowing what the other may or may not have said.

Each witness wrote down statements describing precisely what they had seen. When their independent testimonies were compared, they were remarkably similar: both described a woman dressed from head to toe in white, with shoulder-length light-colored hair. Because both accounts matched and both described some specific details unlikely to be accounted for by mere coincidence, we consider the reports trustworthy. However, when we replayed the recorded video on our monitors—the same monitors on which both team members reported seeing the mysterious woman in white—nothing out of the ordinary was seen.

To this day, we don't know why our witnesses saw something that wasn't recorded on our video, nor do we know the nature or identity of the person they observed. No one was found in the third floor room into which they reported seeing the mysterious woman walk.

Those events were by far the most exciting of the evening, but the show wasn't quite over yet. At 2:45 a.m., we heard the sound of footsteps running on the third floor. Teams dispatched to investigate found nothing. Finally, at 3:21 a.m., we heard the sound of a door creaking through our microphones and headsets. Likewise, teams dispatched to investigate never identified the source.

In terms of weird occurrences and unsolved mysteries, our evening at The McClelland School was one of the highlights of our decades of experience. Very few places put on quite as spectacular a show for us, and we remain baffled by many of the things we witnessed.

References and Further Reading

Purfield, T. (2005). A Child's Place. *The Pueblo Chieftain*. <https://www.chieftain.com/story/news/2005/07/11/a-child-s-place/8470605007/>

8
The Lafayette Vampire:
Lafayette Cemetery

Most of the cases we encounter at the Rocky Mountain Paranormal Research Society involve claims of ghostly activity. Occasionally clients inquiring about private residences will write to us with a claim of something demonic. But when it comes to public venues? The overwhelming majority of claims are of ghosts. That's fine. We like a good ghost story as much as just about anyone else. Still, it's refreshing from time to time to encounter a claim of something *other* than a ghost. We were therefore really excited to learn of the story of the Lafayette Vampire.

The legend comes from a particular plot at the Lafayette Cemetery located at 111 East Baseline Road in Lafayette, Colorado. Supposedly a particular grave there doesn't belong to any ordinary man, but to a vampire.

Figure 8.1. The Lafayette Vampire grave. Photo: Bryan Bonner.

The History

Unfortunately, this is one of those cases in which there simply isn't a ton of history to report. Usually when we work on cases involving public venues (including cemeteries), we find a wealth of information and have to carefully decide which information to omit from our reports because there's simply too much to write about. The historic documentation related to this case, though, is more characteristic of our cases at private residences, wherein we often know very little about the location's history or the biographies of the individuals involved.

With that in mind, this is going to be one of the shorter chapters in this part of the book, but we'll still do our best to provide as much historic context for what follows as we can.

Lafayette Cemetery itself is a municipal cemetery located in Lafayette, Colorado, a part of Boulder County. It was established in 1891 after the land on which it rests was purchased from the Union Pacific Railroad. Be sure you don't confuse it for Lafayette Cemetery No 1, in the Garden District of New Orleans. That cemetery has plenty of ghost stories and paranormal claims of its own (as does so much of the city of New Orleans), but Rocky Mountain Paranormal hasn't made it out there yet.

Though not directly related to the story of the Lafayette Vampire, the cemetery does intersect with a particularly horrific chapter in Colorado history which is worthy of brief note here.

In the 1920s, the Rocky Mountain Fuel Company owned a company town called Serene, Colorado located where the Erie landfill is today to the northeast of Lafayette. It was a small town consisting of little more than company-owned housing, a post office, a coal preparation plant, and the Columbine Mine. On the morning of November 21, 1927, a mining strike turned into a battle between 500 miners and members of the Colorado Rangers, tasked with protecting the mine. The conflict, now remembered as the Columbine Mine Massacre, ended with six people dead. Lafayette Cemetery is the final resting place of five of the six and home to a monument memorializing the event.

We think there's an important lesson here. Much of what we do at Rocky Mountain Paranormal involves using the paranormal stories as a framework to remind people of their history. In many cases, the history feeds directly into the paranormal stories. But as this case shows, even if the paranormal claim (in this case of a vampire) has little documented history behind it, a little bit of digging will reveal something of historic significance not too far away.

But since this chapter is focused on the Lafayette Vampire, it is worth figuring out who the individual in question was and tracing any biographical details possible. Unfortunately, precious little is known.

When one looks at the gravestone in question (see the first page of this chapter), one sees that there are two names associated with the grave.

The inscription on the left side reads:

+ *Romanion*
Trandatir
Born in Par-
Hautibocvina.

This is followed by a line. To the right side of the line we find:

+ *Todor Glava*
Born in Transivania.

Finally, centered below both inscriptions we find the following:

Austro-Ungaria
Died December 1918.

This seems to suggest the presence of two individuals, one named Glava and the other called Trandatir (or Trandafir—likely the "f" was incorrectly transcribed as a "t"). Remarkably, their true identities are known to us, though little else about their lives has been discovered.

With regard to the several nationalities mentioned, we simply need to correct for misspellings on the headstone and understand a little bit of history. The Austro-Hungarian Empire existed from 1867 to 1918. At the time, Transylvania was part of this empire. Also part of this empire was a region known as Bucovina, which encompassed what is now the village of Bucovina (in Romania) as well as the village of Pârhăuti.

The individual on the right—the one mostly associated with the vampire stories—is Theodore "Fodor" Glava. Not much is known about his life except that, as the headstone suggests, he was born in Transylvania (the stone's misspellings notwithstanding). His obituary, published on December 6, 1918 in the *Lafayette Leader*, provides some further insight: "Theodore Glava, an Austrian miner employed at the Simpson mine, died suddenly Wednesday morning, following an attack of influenza. He had so far recovered as to be up town the evening before his death, but suffered a relapse. He was aged 43 years and is survived by his wife, who is in Austria. Burial will be made this afternoon."

Since December 6, 1918 was a Friday and the obituary mentions Mr. Glava died on Wednesday, we can assume his proper date of death was December 4, 1918 and that his burial occurred on December 6 of the same year.

Unfortunately, we don't know much else about his life except for the date and nature of his death, that he was a poor miner (we know he was poor because his grave is located in the cemetery's potter's field and its headstone is marked by the aforementioned misspellings) and that he had a wife who remained somewhere in the Austro-Hungarian Empire. Her identity is unknown to us. One clue is found on his draft card which lists his nearest relative as one Sofich Glava, still residing in Hungary as of the card's registration on September 12, 1918, but it's not clear whether she was his wife or some other kind of relation.

It's a bit harder to conclusively determine who the other grave belongs to. One interpretation is that "trandafir" was the Romanian word for "rose" and so the other inscription might be a reference to Glava's wife. However, this is not the most likely interpretation. Though the translation of the Romanian word "trandafir" to the English word "rose" is correct, a more plausible interpretation can be found in the December 13, 1918 edition of the *Lafayette Leader*, which contains the following obituary for one John Trandafir: "John Trandafir, a native of Rumania, and an employee of the Simpson mine, died of pneumonia at St. Joseph's hospital Wednesday afternoon of last week. Death came a few hours after he reached the hospital. The deceased was aged 27 years and had been in Lafayette about four years. He is survived by his parents and three brothers, who reside in Rumania. He was a member of the Orthodox Greek church, and funeral services were held under the auspices of that church last Sunday afternoon, conducted by the Rev. A. Kaimakan of Denver. Music was furnished by the City Park band of Denver, and the remains were laid to rest in the Lafayette cemetery."

If one follows the dates mentioned in these obituaries, we find that both Theodore Glava and John Trandafir died on December 4, 1918. Given that both men were from the same part of the world, both worked at Simpson Mine, and shared graves weren't considered inappropriate for a potter's field in that era, it seems likely that these are the two men buried beneath the headstone in question. Furthermore, no other Glavas or Trandafirs can be found on any headstones in Lafayette Cemetery.

It's also worth making a historic note that deaths from influenza were quite common in 1918. Indeed, though Trandafir's cause of death was listed in the obituary as pneumonia, it's likely that both men fell victim to the deadly Spanish Flu epidemic of that year.

This pandemic began in March, 1918 at the end of World War I. The first documented case of the virus was Albert Gitchell, an army cook stationed at Camp Funston in Kansas who received his diagnosis on March 4, 1918. Because the war was coming to an end, soldiers were returning to their home countries, and prisoners of war were being released, the virus quickly spread around the world. Contrary to popular opinion, it didn't actually originate in Spain. However, newspapers in countries involved in the war were slow to report on the virus in an effort to boost morale, while the press in Spain (a neutral country in the first World War) reported it openly, resulting in the opinion that it spread more quickly in Spain than elsewhere.

By the end of the pandemic, it had infected over 500 million people and killed, by best estimates, between 17 million and 50 million people worldwide.

That two poor miners living in Lafayette Colorado should have died from the disease on the same day is, unfortunately, not surprising. While most influenzas disproportionately affect the very young, the very old, and the infirm, the Spanish Flu insidiously caused a cytokine storm (an uncontrolled overreaction of the immune system) in the typically-stronger immune systems of younger adults. Poor miners with strong immune systems but working under harsh conditions and without access to quality medical care would have been among the disease's easiest victims to attack.

Nothing in this historic record suggests that either man was (or is) a vampire, so where did the legend come from, and what are the particular claims? We turn our attention to that topic now.

Paranormal Claims

We've been unable to determine the precise origin of the claims that one of the men buried in the grave in question was actually a vampire. We certainly haven't found any accounts (in newspapers, journals, or any other sort of record) of vampiric activity around the time of their deaths in 1918. So, in the absence of those kinds of historic records, why do people think one of these men was a vampire?

It seems that the story has its origin, most likely, in the fact that a large tree stands just below the headstone, seemingly growing out of the grave itself. According to some variations of vampire lore, one can tell that a grave belonged to a vampire—specifically a vampire who'd been killed—if a tree grew from the grave. Such trees are thought to have grown from the wooden stakes used to dispatch the vampire.[1]

Figure 8.2. Tree growing from alleged vampire grave. Photo: Bryan Bonner.

In addition to the idea that the tree over the grave grew from the stake through the vampire's heart, a variety of paranormal occurrences are reported when people visit the grave.

Batteries in electronic equipment are meant to drain when in proximity to the grave, even if they're not being used.

People report seeing a "shadow figure" standing at the grave. Others describe

---

1   A remarkably similar story concerns a large tree growing from a grave in a cemetery called El Panteon de Belen in Guadalajara, Mexico. As is the case with the Lafayette Vampire, legend claims that the tree grew from the wooden stake the townspeople used to kill the vampire.

having heard disembodied voices or seeing strange lights at the gravesite. Some believe that if one visits the grave at midnight[2], you can see the vampire sitting on top of the gravestone. When locals (presumably teenagers) have dared each other to stand near the grave, some have reported seeing a tall skinny man with long fingernails and wearing a dark coat sitting on the grave.

One story has it that a rose bush on the grave started growing from the vampire's fingernails. That "trandafir" is the Romanian word for "rose" makes this a particularly creepy story, we think.

Finally, there's a legend that people walking through the cemetery alone find themselves beaten up. When police arrive to investigate the batteries, however, they find only one set of footprints, leading back to the grave.

Our Investigation

It's hard to know how to investigate a vampire. It's not exactly something one can summon on demand. And if the vampire is already deceased, as many versions of the story attest, there's really not a whole lot we can do to prove the claim true or false. However, we still do our best to separate fact from fiction.

To begin with, we visited the grave and documented what we observed. First of all, there absolutely is a tree growing from where the grave ought to be, just a couple feet "below" the headstone. We did not, however, see any rose bushes growing on or near the grave. On some visits, we did observe cut roses placed on the grave by some visitor, but that seems unrelated to the vampire lore.

Does the presence of a tree growing from a grave indicate the presence of a vampire? It's hard to prove a negative, but we can think of a few reasons why this seems unlikely.

Probably the most important reason is that the phenomenon defies botanical reality. Wooden stakes are carved from tree material, yes, but they are not living seeds. They're formed from dead wood. If one buries a wooden stake in the ground, it does not grow into a tree. Even if one buries a wooden stake in the ground embedded within some nutrient-rich material like a heart, it does not grow into a tree.[3]

2   Please don't sneak into cemeteries after hours. If you have legitimate business there in the middle of the night (as we sometimes do when investigating paranormal claims), the appropriate course of action is to explain yourself and get proper permission from the cemetery's management rather than sneaking around after hours which can put the grave property, your safety, and any guards' safety in danger. We wish cemeteries were open twenty-four hours just as much as you do, but that's not an excuse for trespassing.

3   Just for the fun of it, one of our members once buried a wooden vampire stake embedded in a chicken heart in his back yard. It did not grow a new tree.

Furthermore, if a tree growing from a grave *does* indicate that the grave belonged to a vampire, that would mean there are a lot more vampires running around than anyone would be comfortable with. While most graves are, of course, unadorned by trees, the sight of trees growing from graves is not all that uncommon, particularly in older cemeteries. Even withing Lafayette Cemetery, we can find several other graves with trees growing from them, and none of these (to the best of our knowledge) are associated with any vampire lore.

Figure 8.3. A (non-vampire) grave with a tree. Photo: Bryan Bonner.

This story also contradicts some of the paranormal claims made regarding the grave. If the tree grew from the wooden stake that killed the vampire, then the vampire is, by definition, dead. Who, then, is the mysterious figure who is meant to be seen around the grave or occasionally to batter individuals walking through the cemetery? Unless one assumes the figure is the ghost of the vampire (which seems like a bit of a stretch even in paranormal lore), these variants of the story seem to undermine each other.

None of this stopped us from doing our due diligence, though. We spent some time monitoring the gravesite, and didn't observe any anomalous behavior. The batteries in our devices did not drain. No EMF anomalies were reported. No mysterious figures showed up. In fact, we didn't see or hear anything at all out of the ordinary.

It seems most likely to us that the origin of this story comes from the fact that one of the men buried in this grave came from Transylvania. Ever since Bram Stoker published his brilliant and timeless novel *Dracula*—and certainly in the decades since Hollywood has capitalized on vampire lore—the Romanian region has been almost synonymous with vampire lore. Our suspicion is that someone saw the tree growing from the grave, noticed the man buried there was Transylvanian, and put those two pieces together.

In the decades since, we've noticed that most of the stories associated with the Lafayette Vampire aren't coming from locals visiting the cemetery but rather from other ghost hunting or paranormal investigation groups. It seems that these individuals keep building the mythology off of each other's reports, almost without regard to the fact that, though locals have *heard of* the vampire legend, we haven't found many outside of the paranormal groups who claim to have actually seen or experienced anything.

Honestly, we like this story on some level. As we mentioned in the introduction to this chapter, it's nice to get a bit of variety in our case files rather than just looking for ghosts. In fact, this represents our only vampire case to date. But on another level, it seems a bit distasteful that so many people have associated two innocent men—presumably hardworking immigrants—with vampiric evil on the basis of no real evidence at all. That doesn't seem like the legacy these men deserved. On the other hand, there's something nice about the idea that the vampire story has given us cause to remember these men (at least insofar as we've been able to learn a few minor details about their lives) more than a hundred years later. It may not be the legacy they deserved, but at least it's an excuse to remember the dead, and that seems to be worth something.

Incidentally, the Lafayette Vampire somehow managed to transcend mere local legend and has been featured in several fictional stories.

One such reference turned up when we stumbled across a campaign (now deleted, as far as we've been able to tell) from an online vampire roleplaying game and noticed that not only did the Lafayette Vampire make an appearance, but so did Rocky Mountain Paranormal, in fictionalized form.

In 2010, writer/director Nicholas Bernhard produced a silent short film centered on the Vampire entitled *Glava*.[4] According to its plot summary on IMDb (the Internet Movie Database): "On Halloween night, for [sic] friends make their yearly pilgrimage to the local cemetery. The next day, one of them is found dead. The remaining friends vow to discover the connection between his murder and the urban myth that haunts the cemetery: that of the poor Romanian coal miner with vampiric tendencies."

Finally, the vampire, under the name of Fodor Glava, makes an appearance in the 2012 novel *The Dead of Winter* by Lee Collins. The publisher's description of the novel reads: "Cora and her husband hunt things—things that shouldn't exist. When the marshal of Leadville, Colorado comes across a pair of mysterious deaths, he turns to Cora to find the creature responsible, but if Cora is to overcome the unnatural tide threatening to consume the small town, she must first confront her own tragic past as well as her present."[5]

The Lafayette Vampire may not have turned out to be much of a vampire at all, but the urban legend has given us some good stories and an excuse to visit an old cemetery and remember the people from our past, and that's always a win in our book.

---

4   As of this writing, the (shorter) re-edited fifth anniversary edition of the film is available for viewing on YouTube at https://youtu.be/gBX64DbLWuU.

5   Collins, L. (2012). *The Dead of Winter*. Watkins Media Limited.

References and Further Reading

Collins, L. (2012). *The Dead of Winter*. Watkins Media Limited.

Getz, C. O. (2010). *Weird Colorado: Your Travel Guide to Colorado's Local Legends and Best Kept Secrets*. New York: Sterling.

9
Murder at the High School: Platte Canyon High School

We all love a good ghost story or urban legend. But while we enjoy the tales, it's important to remain grounded in reality. Loss of that grounding can result in the worst of all possible consequences. The story of Platte Canyon High School is a constant reminder of the dark side of paranormal lore, when the stories get taken too far.

Platte Canyon High School, located at 57243 US Highway 285 in the small town of Bailey, Colorado, serves approximately 300 students at any given time. It's also the site of an urban legend surrounding an assault and suicide on the school grounds, which has led to numerous claims of paranormal activity.

Unfortunately, after our investigation, things took an even darker turn....

Figure 9.1. Platte Canyon High School. Photo: Bryan Bonner.

The History

There's remarkably little history regarding Platte Canyon High School. Or at least there was remarkably little history before our investigation. The most noteworthy event in the school's history occurred after our visit and we'll discuss it later, to maintain the proper chronological presentation of the story.

Up until that time, the history of Platte Canyon High School was remarkably similar to the history of any other small town high school. Even the town in which it resides feels very much like any other American small town in terms of its history.

That history began in 1864 when William Bailey settled a ranch about thirty miles southwest of Denver and called it Bailey's Ranch. He eventually built a hotel at the ranch and the town, which eventually became known as Bailey after its founder and his ranch, began to grow. By 1878, the Denver and South Park Railroad reached the small town, and a post office was established on November 20, 1878.

In addition to the usual homes, shops, and businesses, the town features a number of historic locations and attractions including the Coney Island Hot Dog Stand (a building in the shape of a hot dog) which was established in 2006 and, of likely interest to readers of this book, The Sasquatch Outpost, a gift shop specializing in Bigfoot memorabilia.

And of course, it also includes the Platte Canyon High School, which was established in 1957. An annex was added in the year 2000, but most of the school's history consists merely of business as usual.

Except for one urban legend.

The Legend

Let us make it abundantly clear right from the start (even though we may be spoiling a bit of our investigation report later) that we have not been able to conclusively verify this legend. Despite years (going on two decades at this point) of searching, we have yet to find a police report, a newspaper article, or any other official or trustworthy record that provides an account of the real story. However, we'll present the urban legend for what it's worth because it's an important piece of the Platte Canyon High School lore.

There is a storage area attached to the school gymnasium. According to the legend, several of the students once went to this room to retrieve something (or perhaps to place something in storage), whereupon they discovered the body of a faculty member who'd hanged himself.

Figure 9.2. Storage room in Platte Canyon High School. Photo: Bryan Bonner.

As if that discovery wasn't gruesome enough, it was later discovered that he'd assaulted a fourteen-year-old girl (a freshman at the high school) immediately before his suicide. Apparently he'd held her hostage in that storage closet for an unspecified time before finally deciding to end his own life. Neither the perpetrator's nor the victim's identity are known.

Students, faculty, and members of the local community are very well aware of this legend, as are paranormal buffs. It seems that, in the aftermath of this tragic event, the school has become home to some paranormal activity.

Paranormal Claims

It's not difficult to see how an urban legend like the one described above could easily turn into a paranormal claim. Any violent death or suicide runs the risk of turning into a ghost story. When one involves a young student and a faculty member of a school, and actually takes place on the school grounds, it seems almost inevitable that people will begin to tell stories about ghosts. As we've discussed before, schools can be naturally creepy places, so combining that natural uncanny feeling with a truly horrific story leads to quite the ghostly tale.

Among the people who've reported paranormal activity was one of the janitors working in the school at the time of our investigation. He told us he'd repeatedly seen things moving out of the corner of his eye.

Another school employee told us of an eerie experience in the school's "B" Gym. After cleaning the area after hours, a faculty member returned to the room and found a hairbrush placed in the middle of the gym floor along with several other items that had just been put away. No hairbrush was previously in the room.

Students and employees have regularly reported seeing or hearing some fairly standard paranormal phenomena: shapes of people moving, moaning sounds, footsteps, and voices coming from the particular storage room in which the legend says the assault and suicide took place.

Our Investigation

We were approached by Platte Canyon High School staff with a proposal to conduct an investigation into the strange occurrences several of their students and colleagues had been witnessing. Their idea was for several students in the school who'd formed a paranormal club of their own to join us on the investigation as a kind of hands-on after-hours learning experience. Because we're all about education, we jumped at the opportunity and arranged a schedule for the investigation.

The first time we showed up, the teacher we'd contacted was a no-show, so we retreated to a local diner and asked the locals about the legend. Several high school students were at the restaurant at that time. Every single one of them either knew of the legend or had a story of some unexplained phenomenon. We were intrigued.

While at the restaurant, we were contacted by a student's mother who said she would meet us at the school for our investigation, and we made the necessary arrangements. We arrived on October 29, 2005, ready to go. Unfortunately most of the people—faculty and students alike—who were supposed to join us on the investigation had flaked out, but we nevertheless proceeded with a small group of curious minds. We were joined for the evening by one parent and three students.

Before arriving at the school, however, we'd done our due diligence regarding the school's history and the urban legend. Unfortunately, this was one case in which we came up very nearly empty. That's uncharacteristic for us because we're a stubborn lot and we like to keep pressing until we solve our mysteries, but this one just continues to elude us.

Normally, we'd assume that lack of evidence suggests the legend is just a made-up story. It's the sort of event one would expect to be a matter of some public record somewhere, so the absence of evidence would seem, if not conclusive, at least suggestive that the story was false. However, the matter is a bit more complicated than that. Over the years, we've spoken to numerous sources off the record who told us that the urban legend was based on some kind of real event but that they couldn't elaborate. Further information has proved quite elusive.

So honestly, we have no idea whether the legend is true or not. If it is, we also have no idea why the records have been so well-hidden. Perhaps, if the story is true, someone connected to either the perpetrator or the victim wanted the story kept quiet and had sufficient influence to keep it that way. On the other hand, perhaps the story is merely a fabrication and the people we spoke to merely wanted to keep the mystery alive for whatever reason. On the other hand (that would be three hands for those of you keeping count), perhaps there was a real event but it bore such little resemblance to the legend that people just don't remember it.

Regardless, though our historic research was uncharacteristically unfruitful, we still had an on-site investigation to conduct. We began setting up our equipment at approximately 10:00 p.m. and packed up and went home at about 4:00 a.m. However, this evening included the clock change for Daylight Savings Time, so the duration of our investigation was actually an hour longer than those times would suggest, for a total of seven hours.

Things took a little longer to set up than usual because we were also instructing the students while we worked. Even still, we had things ready to go shortly after arriving. Most of the monitoring equipment kept watch over the storage area, though we also took account of the surrounding rooms. At the beginning of the evening, we took baseline EMF and temperature readings of the storage room, the A and B gyms, and the pool area.

Just as we finished setting up the equipment, we stopped for a moment because several of our company reported hearing noises that sounded like footsteps coming from the storage area. Six people heard the noise. Though we can hypothesize potential natural explanations, the source of the sound remains unknown.

Throughout the evening, as is our way, we conducted hourly EMF and temperature readings at several locations. Interestingly, we actually recorded some anomalous readings during these measurements. At the entrance to the B Gym and near the back wall in the A Gym, apparently at random intervals, the EMF readings would increase by about 1.5 to 4 milligauss for durations ranging from five to thirty seconds. No other activity (paranormal or otherwise) was noticed at the same time as these readings, but we were not able to find any cause for the anomaly in the area. The source of these readings remains unknown to us.

As the evening wore on, temperatures did decrease. However, this is normal. The investigation took place in October, so the declining temperatures were consistent with decreasing outdoor temperatures in the autumn season.

Our protocol for monitoring the attic involved a combination of remote monitoring with video and audio recording as well as in-person monitoring. Equipment included two infrared cameras, two digital audio recorders recording directly to CD, thermometers keeping track of temperature, and two infrared lights. This was combined with hourly EMF readings. While most of the team watched remotely on screens, shifts of two people at a time would sit silently in the storage area, as motionless as possible, to see if they could see anything while being careful not to interfere with the data. This is a common practice for us. Because some paranormal lore suggests that activity is more likely in the presence of a human observer than merely on video alone, we've developed a practice in which individuals who need to rest for a while get to be "bait." That is, they go to the (allegedly) most haunted location and rest in silence while the rest of the team watches remotely to see what happens.

Throughout the evening in question, several of the individuals baiting the attic area reported some minor anomalies. At one point, two individuals both reported the sound of something moving in the room as well as the sight of a light in the corner. Nothing was seen on the video recording, however.

As these small anomalies continued, we noticed an interesting pattern. Activity seemed to increase when particular individuals were present in the attic—including one of our team members as well as all three students. However, when certain other individuals were present in the attic, anomalous activity ceased. If the activity had only increased in the presence of the students and ceased whenever any of our investigators were present, we might have suspected the students of playing a prank on us. However, one of the individuals whose presence seemed to prompt the anomalous activity was one of our team members. Plus, all individuals in the attic were constantly watched on video, so pranks don't seem like a likely explanation, and we haven't figured out any further information as of yet.

When we wrapped up, we considered it a fairly run of the mill investigation. We were certainly glad to be able to work with the students, and hopefully they found it to be an enriching experience, but we didn't find anything that got us too excited. There was just enough anomalous activity to give us something to think about, but nothing that would really make the case stand out. The same was true with regard to the historic research concerning the urban legend. We found just enough to pique our interest, and not nearly enough to actually learn anything useful or conclusive. We left feeling like it would go into our records as just another case among many.

About a year later, our phones started ringing off the hook. Something horrible had happened at Platte Canyon High School.

The Hostage Crisis and Murder

At 11:40 a.m. on September 27, 2006 (just less than a year after our investigation), 53-year-old Duane Roger Morrison entered Platte Canyon High School carrying a backpack. He told school staff that the bag contained explosives and demanded they follow his instructions. Allegedly he discharged a firearm as a warning shot when one teacher failed to comply.

He entered a classroom and told all of the male students to leave, leaving him with six female students as hostages. As the other students evacuated the school and law enforcement officers began to arrive on the scene, Morrison is said to have sexually assaulted the six female students. The precise nature of these assaults has not been made a matter of public record.

When law enforcement arrived, negotiations began initially by telephone. Morrison's demands were never made clear, except that he wanted the police to back away.

The police, of course, negotiated in the interest of getting the hostages released. Because Morrison did not particularly care to speak to officials directly, he began releasing some of the hostages, using them to relay messages to law enforcement officers outside.

Once four of the six hostages had been released, negotiations continued to intensify. The feeling among law enforcement officers was that things were about to reach a critical point.

During this time, 16-year-old Emily Keyes, one of the two remaining hostages, managed to send a brief text communication to her father, John-Michael Keyes. He'd received word that there'd been an incident at his daughter's school and sent the text message "R U OK?" to Emily. She responded "I love u guys." Her father's follow-up message, "Where are you?" received no reply.

In his indirect communications with the police, Morrison had indicated he would stop negotiating at 4:00 p.m. With two hostages remaining and time running out, the police feared what would happen if Morrison's deadline was allowed to pass. The decision was made to breach the classroom.

At 3:45 p.m., Jefferson County SWAT officers used explosives to burst through the door in an attempt to rescue the hostages. They exchanged fire with Morrison, who attempted to use the two girls as human shields. When 16-year-old junior Emily Keyes attempted to run, Morrison shot her in the head.

As police returned fire, non-lethally wounding Morrison, the other hostage escaped. At this point, Morrison turned his firearm on himself and ended his own life.

The gunman died on the scene. Young Emily was rushed by helicopter to St. Anthony's Central Hospital (now known as St. Anthony's Hospital) in Lakewood, where she was pronounced dead at 4:32 p.m. following emergency surgery.

Morrison's motivations were never made entirely clear. Some reports suggest that he targeted specific students, but even at that the true motivation behind his actions died with him. He left a rambling and barely coherent fifteen-page suicide note in which he describes having a tormented relationship with his father, a random grievance with a motorcycle dealer, and instructions for the disposition of his meager belongings, but no clue as to the thinking behind his actions.

In the past, he'd been a petty criminal with a 2006 arrest for obstructing police and a 1973 record for larceny and marijuana possession, but nothing of the caliber of his final act.

Though his last known address was in Denver, it appears he'd been homeless and living in his automobile for some time prior to the attack.

If that was all we knew about Morrison, that may have been the end of it. However, the press got wind of the fact that he'd been employed by the company that operated Primitive Fear, a seasonal haunted house in the Denver area. And then they heard that Rocky Mountain Paranormal had been at the school about a year earlier.

In some circles, a picture began to emerge of a man obsessed with ghost stories who tried to recreate the urban legend he'd read about. However, we emphasize that this was mere speculation. The reality is, Duane Roger Morrison was a worthless and bitter man whose anger at the world resulted in a senseless tragedy for no other reason than he apparently sought revenge not against an individual but against existence itself and enacted it in one of the most cowardly ways possible.

If it is true that his actions stemmed from an obsession with ghost stories (and we don't think it is; there's no evidence that he even knew of the legends at Platte Canyon High School), then consider it a lesson regarding the importance of enjoying these tales while remaining grounded in the real world. If it's not true, consider it a lesson in how superficial similarities between a legend and a real event can cause people to misinterpret events and motivations in an unproductive way.

Regardless, we choose not to end this chapter by discussing a worthless and destructive poor excuse for a human being. Instead, we prefer to focus our attention toward the memory of the victim, 16-year-old Emily Keyes, whose father described her as a "smart, fiery, beautiful teenage girl." Following the tragedy, Emily's parents founded the I Love U Guys Foundation in her memory.

10
A Place of Death and Despair:
The Denver County Jail

Any building can be the site of a ghost story, regardless of its history or whether any dark events happened in its past. However, the darker and more violent a building's history, the more likely it seems it will be the subject of ghost stories. With that in mind, few places seem riper for allegations of hauntings than the Denver County Jail. Like any prison, it's associated with many of the darkest and most horrifying events that have happened in its region.

It's also the site of one of our more interesting investigations. Though the jail, located at 10500 East Smith Road in Denver, is still active and operated by the Denver County Sheriff's Office, we were able to conduct an investigation into some of the complex's many buildings during a time when inmates had been relocated pending demolition of certain buildings as part of a renovation project in 2010.

Figure 10.1. The Denver County Jail. Photo: Bryan Bonner.

The History

The first thing that could be a matter of some confusion when speaking of "Denver Jail" is that there are actually two jails in Denver. The first is the Denver County Jail, which is used for the long-term housing of convicted inmates. The second is the Van Cise-Simoet Detention Center, more commonly called simply the Denver Detention Center. This second location is used primarily for the short-term housing and processing of inmates as they move through the criminal justice system. While local residents are more likely to have direct contact with the second facility (perhaps when posting bond for someone in their social circle), the first is the location of our investigation.

Cities need jails. It's an unfortunate truth of the imperfect world in which we live (after all, as James Madison wrote in *The Federalist No. 51*, "If men were angels, no government would be necessary"). Therefore, the history of the Denver Jail extends almost as far back as the founding of the city itself. The old Denver County Jail was a rock ashlar building erected in 1891 at the corner of Colfax and Kalamath in the heart of Denver. However, this building was only used until 1956 and eventually demolished in 1963.

The current Denver County Jail, and the subject of this chapter, opened for "business" the same year the previous location closed. It was designed to house as many as 1,500 inmates, consisting both of long-term inmates convicted of misdemeanors as well as those felons awaiting transfer to the Colorado Department of Corrections.

Denver County Jail is not a single building, but a large campus. At the time of our investigation, it consisted of a primary structure of 18 connected buildings as well as 5 separate buildings including buildings for staff and Mechanical Building 21 which housed the boilers and other utilities. Heating and utilities for the entirety of the campus are provided through service tunnels located under all of the facilities.

By 2010, the facility designed to house up to 1,500 inmates had a daily average population of 1,986. Renovations, necessary to house the growing population, were desperately needed, and the plans were put in place to demolish several buildings to make way for newer, improved structures. Buildings 6 through 13 were demolished as part of this project and replaced with the current facility as it stands as of this writing. The facility's current population as of this writing, having declined somewhat in recent years, is approximately 1,700 individuals.

There's not a lot of terribly interesting history regarding the building itself. The real history of Denver County Jail is the history of its inmates. Of course, we cannot even begin to provide complete histories of all of the individuals who've passed through the facility. Even if we disregard petty criminals or individuals with relatively uninteresting life stories, there are still far too many to discuss in this book (and that's to say nothing of potential privacy concerns). However, there are several noteworthy inmates and incidents in the jail's history that are worthy of brief mention.

One such inmate was John "Jack" Gilbert Graham, a notorious mass murderer in the 1950s. On the evening of November 1, 1955, United Air Lines Flight 629 was passing over Longmont, Colorado when an explosion destroyed the plane. Law enforcement officials were quickly able to determine that Graham had planted a dynamite bomb in the checked luggage in an attempt to kill his mother, both in revenge for her having left him in an orphanage as a child and to collect the $37,500 (equivalent to about $417,000 in 2022) life insurance policy he'd taken out immediately before the flight's departure. As for the other passengers aboard? Graham didn't care. After his capture, he explained to a psychiatrist, "The number of people to be killed made no difference to me. It could have been a thousand. When their time comes, there is nothing they can do about it."

The bombing resulted in the deaths of all passengers and crew aboard the plane: a total of 44 individuals (45 if you count the unborn child of victim Carol Bynum, who was pregnant at the time of the crime) whose ages ranged from 13 months to 81 years.

After the explosion, the plane crashed in farmland east of Longmont, with debris spread over about five miles. Local farmers assisted first responders in attempting to find and rescue survivors, but there were none. A police officer's haunting report from the scene says it all: "No ambulances are necessary." Even decades later, the locals who witnessed the wreckage find reminders of the terrible event. Some have reported seeing ghosts.[1] Others describe finding small personal items like eyeglasses on their farms years later.

Graham was tried for murder in Colorado's first ever televised trial. Interestingly, he was not tried for all 44 counts of murder but instead only for a single count of first degree murder for his mother, Daisie King.[2] Nevertheless, he was quickly sentenced to death and housed at the Denver County Jail until his execution in the Colorado State Penitentiary gas chamber on January 11, 1957.

In July of 2022, Marian Hobgood Poeppelmeyer, daughter of victim Marion Pierce Hobgood, asked the city of Longmont to honor the victims with a memorial. As of this writing, the city hasn't responded, but Rocky Mountain Paranormal, for our part, will honor the victims here.

---

1   We have not (yet) investigated these claims.

2   This was not an oversight. The prosecutor on the case decided that a charge of premeditated murder of a single person would result in a much faster trial and higher probability of conviction than a more complicated trial involving dozens of other charges.

The victims of Flight 629, in alphabetical order, were:

- Fay E. "Jack" Ambrose, 38
- Samuel F. Arthur, 38 (flight engineer)
- Bror H. Beckstrom, 48
- Irene Beckstrom, 44
- John P. Bommelyn, 53
- Frank M. Brennan, Jr., 36
- Barbara J. Bruse, 23
- Louise D. Bunch, 61
- Carol Bynum, 22 (and unborn child)
- Horace Brad Bynum, 32
- Thomas L. Crouch, 23
- Carl F. Deist, 53
- John P. Des Jardins, 42
- James Dorey, 58
- Sarah Dorey, 55
- Elizabeth D. Edwards, 57
- Gurney Edwards, 58
- Helen Fitzpatrick, 42
- James Fitzpatrick II, 13 months
- Lee H. Hall, 38 (pilot)
- Goldie Herman, 59
- Vernal Virgil Herman, 69
- Elton B. Hickok, 40
- Jacqueline L. Hinds, 26 (stewardess)
- Marion P. Hobgood, 31
- John W. Jungels, 57
- Daisie E. King, 53 (the killer's mother)
- Gerald G. Lipke, 38
- Patricia Lipke, 36
- Lela McLain, 81
- Frederick Stuart Morgan, 48
- Suzanne F. Morgan, 40
- Peggy Ann Peddicord, 22 (stewardess)
- James W. Purvis, 45
- Herbert G. Robertson, 43
- Dr. Harold R. Sandstead, 50
- Sally Ann Scofield, 24
- Jesse T. Sizemore, 24
- James E. Straud, 51
- Clarence W. Todd, 43
- Minnie Van Valin, 62
- Dr. Ralph Waldo Van Valin, 72
- Donald A. White, 26 (copilot)
- Alma L. Winsor, 48

Unfortunately, the Denver County Jail was also the birthplace of a prison gang whose activities have spread outside of the jail's walls. In 1995, inmate Benjamin Davis (a white man) was beaten and nearly killed by a black inmate with a sock stuffed with soap bars. He responded by forming 211 Crew, a white gang. Initially, the goal was to create the illusion of a white gang to help protect its members from the black and Latino gangs in the jail. However, as membership grew and began recruiting, 211 Crew morphed into a more active "real" gang and has been associated (somewhat controversially) with white supremacist ideology.

It's an unfortunate reality that, as prison gangs grow in membership and some of those members are eventually released, their violence begins to spread outside of the prison walls. Such was the case with 211 Crew.

Oumar Dia was an immigrant from a village called Diorbivol in Senegal. By all accounts, he was a kind, hardworking, upright citizen. He worked long hours cleaning rooms and running luggage at Denver's Hyatt hotel. Apparently he was not selfish with his money. He sent the bulk of his paychecks back home to Senegal to help his wife and three children. He also reportedly made regular donations to charitable causes.

On November 18, 1997, Dia was standing at a downtown Denver bus stop waiting for his bus when he was approached and shot by 211 Crew members Nathan Thill and Jeremiah Barnum. Witnesses described the attack as apparently random and perhaps racially motivated. Dia was shot three times and killed. A witness, Jeannie VanVelkenburg was also shot and paralyzed.

Probably the most notorious action of the 211 crew occurred several years later. In March of 2013, gang member Evan Ebel told another member that he was "taking care of things on the street for other inmates" and would "likely die soon." Seven days later, his plot, which many believe was not a solo action but part of a larger conspiracy planned by the 211 Crew gang, began.

On March 17, 2013, Ebel attacked part-time Domino's pizza delivery driver and father of three Nathan Leon. He forced the pizza man, at gunpoint, to read a rambling statement that condemned prison officials specifically for the practice of holding inmates in solitary confinement. As soon as the message was recorded, he shot and killed Leon and stole his pizza delivery uniform to use as a disguise.

Two days later, wearing the pizza delivery disguise, he arrived at the home of Colorado Department of Corrections head Tom Clements. According to Clements' wife, Lisa, she and her husband had been watching television when someone (we now know to be Ebel) knocked on the door. Clements answered and was immediately shot twice, the force of the attack sending him backward and down the steps. He survived long enough to shout for his wife to call 911 but was pronounced dead by paramedics half an hour later. Ebel fled the scene and ended up in Texas. Law enforcement in the Lone Star State, however, recognized his vehicle. A high-speed chase resulted, during which Ebel crashed his car and was shot and killed in the ensuing shootout with police.

Because life is often stranger than fiction, Ebel's father, Jack Ebel, was an old friend of then-Governor of Colorado John Hickenlooper, the very same man who'd appointed Clements to his position in the first place. When Governor Hickenlooper interviewed Clements for the position, they'd discussed the son of a friend who'd been housed in solitary confinement. That individual turned out to be Evan Ebel.

In addition to housing some truly despicable and infamous inmates, some of whose violence unfortunately spilled into the wider community, Denver County Jail itself has been the scene of a number of incidents. Though we don't pretend to present a complete history of all incidents at the institution, we have selected some that we think may be of particular interest. We present these in chronological order.

Returning to the aforementioned case of John Graham, an incident occurred during his incarceration, on the evening of February 10, 1956, soon after both his arrest and the jail's opening at its current location. Though he'd been described as a model prisoner up to that point, at about 5:30 p.m., a deputy sheriff[3] was summoned to his cell by the sound of heavy breathing. He discovered Graham slumped on the floor with his own socks twisted tightly around his neck. A piece of rolled cardboard had been added to make the garrote more effective. The Deputy was able to loosen the makeshift suicide device and call for a doctor. Graham was sedated and restrained in a straitjacket that evening and kept under psychiatric observation for the duration of his incarceration.

Despite being highly secure buildings, occasionally attempts are made to escape jails or prisons. One such attempt occurred at the Denver County Jail on Thursday October 16, 1958. Seventy inmates, including at least one confessed murderer—one Donald Zorens, who had admitted to murdering Denver Patrolman Donald Seick—participated in the riot. Over 100 officers armed with both lethal and non-lethal weapons responded to the uprising. Four officers received minor injuries, but the event was ultimately suppressed without any severe injuries or deaths. At the height of the riot at around 8:00 p.m., 36 of the available 44 patrol cars in the police department had surrounded the facility. Fire trucks also responded, using their spotlights to illuminate the building.

Though no severe injuries were reported, the cell block was left a mess. Showers had been left on, flooding the floors and mixing water with what was left of the evening meal. Mattresses had been overturned, ripped, and scattered. Brooms and brushes were broken. Tables were overturned, and windows smashed. One staff member responded to the riot by saying, "If they think things have been tough, they haven't seen anything yet. I'm taking away their mattresses and radios. There's going to be no more picture shows, and they can eat in their cells."

This riot occurred in the Jail's maximum security cell block, housing 103 of the Jail's 738 (at that time) prisoners. Among those, and thought to be one of the ringleaders of the riot, was James A. Gonzalez, held on charges of armed robbery.

Either late in the evening of August 6 or early in the morning of August 7, 1974, Deputy Sheriff John Derek Osborne was beginning his nightly rounds when two inmates attacked another deputy sheriff. The two inmates overpowered the deputy, stabbed him, and took his keys. In their attempted escape, they came across and attacked Osborne and a third deputy. While the other deputy was able to escape to safety in the "cage" (a secured room for jail personnel), the inmates cut Osborne's throat with their four-inch homemade blade. Osborne, aged 31 years, died of his injuries after being rushed to Denver General Hospital. His two assailants were unsuccessful in their escape.

An inmate named Miguel A. Richardson was convicted of the murder of John

---

3   One of the things we learned during this investigation is that the officers at the jail don't like to be called "guards." They are, rather, sworn deputy sheriffs. For ease of reading and clarity, we will still occasionally use the word "guard" in this book when not referring to a specific individual.

G. Ebbert, one of two security guards (the other guard was Howard Powers) he shot in 1979. In March of that year, Richardson was attempting to rob a motel room in San Antonio, Texas. A guest complained to management and the two guards were dispatched to investigate, whereupon they found Richardson attempting to gain entry into one of the rooms. They apprehended him for a moment, but a gun fell from his waistband. Richardson managed to retrieve the weapon, subdue the guards with handcuffs, then shoot both of them, escaping with their money.

Richardson remained at large for more than a year but was eventually caught and arrested in Denver on June 19, 1980. At this point, his story intersects with ours, as he was held at the Denver County Jail pending extradition to Texas. While confined awaiting extradition, his violent streak continued within the walls of the Jail. On April 20, 1989, after being informed he was to be placed in isolation, he attacked the Jail's Shift Commander and attempted to beat him (presumably fatally) with a fire extinguisher. His attack failed. However, on June 20 of the same year, he was taken to Denver General Hospital for medical care. During the course of his treatment, his handcuffs were removed—a substantial mistake—and Richardson saw his opportunity to escape. Using a shank fashioned from a spike, he stabbed a deputy in the neck and attempted to gain control of the officer's firearm. Fortunately, the quick-thinking officer managed to prevent the firearm from discharging by jamming the loose skin between his fingers and thumb in the space between the weapon's firing pin and the round in the chamber. The shank, which penetrated the right side of his neck all the way to his spine, left him with a hoarse voice and his defensive wounds left him with a disfigured hand, but he survived the attack.

Despite his escape attempts, Richardson was successfully extradited to Texas, where he continued to attack people in failed escape attempts. He was executed by lethal injection on June 26, 2001.

By June 14, 1990, overcrowding was already becoming an issue at the Jail. Officials realized that fewer than 60 beds remained. In order to avoid a prisoner overflow problem, 23 inmates were released early. Though the overpopulation issue would resurface (and would eventually result in a substantial renovation of the facility), the crisis was averted for a time.

On June 1, 1999, an accused armed robber successfully escaped from the Jail. Apparently he accomplished this rare feat merely by walking out with a group of visitors. Without condoning crime in any form, one almost has to admire the chutzpah.

On Christmas Eve of the year 2000, inmates settled in to watch a screening of *The Cell*, a science fiction thriller which had come out earlier that year. During the showing, however, a VCR broke down, resulting in a prison "rampage."

In 2002, an inmate suspected of serial rape was stabbed by another inmate. The attacker was supposed to be on lockdown but instead somehow managed to hide in a shower. When the victim walked by with two deputies, he attacked. The victim was treated at Denver Health Medical Center for stab wounds to the left arm, leg, and ankle, but was not permanently injured or killed and was returned to jail soon thereafter.

Not all prison deaths are the result of murders. October 24, 2004 saw the death of one inmate who had hidden a package of cocaine in a body cavity. When the package ruptured, the inmate collapsed, suffered a seizure, and died of an accidental overdose before medical personnel could respond.

Two days later, a maximum-security inmate was stabbed repeatedly in the upper torso by a fellow inmate using (as is so often the case in these attacks) a homemade shank. The injured prisoner was taken to the hospital and made a full recovery.

Several other incidents have come to our attention for which we don't have precise dates. They're nevertheless interesting, so we present them (however briefly) now.

While being moved to his cell, one prisoner managed to free himself from his escorts. He ran up the stairs to the rear of the third floor tier and jumped to his death. What was particularly surprising about this incident was that he was being held on charges "only" of destruction of private property and threats. Typically, one wouldn't predict a suicide to result from an impending jail sentence related to such crimes. Suicides would be more expected if one faced a life imprisonment or something similar. However, it was later discovered that this inmate's lover was being housed at the same facility and he took his own life in an ill-considered display of affection. Five months later, the lover attempted to do the same thing but was stopped by a fellow inmate.

Figure 10.2. Location of suicide jump. Photo: Bryan Bonner.

One of the suicides that's more in line with what we expect was that of someone who faced charges of sexual assault and kidnapping. He hanged himself in his cell with a bed sheet. Because he hanged himself on the lower bunk of his bed, passing officers initially thought he was just sitting at the end of the bed, so his suicide was not discovered for approximately five hours.

Finally, an inmate was being held on charges of menacing with a deadly weapon. Unfortunately, he was housed with the wrong cellmate. The latter stabbed the former in his cell. He died on the way to the hospital.

Clearly, as the title of this chapter indicates, the Denver County Jail is a place of death and despair. That's not an indictment of the Jail or its staff. We're not saying this particular jail is any better or worse than any other. Nor are we making any political statement one way or the other about criminal justice policy. Rather, we're simply observing that jails and prisons are repositories of society's guilt and shame. As such, it's unfortunately to be expected that bad things will happen there from time to time. And as we've already established, where bad things happen, ghost stories are almost certain to follow.

Paranormal Claims

Both inmates and staff of the Denver County Jail have reported a variety of unusual phenomena over the years. One of the challenges in pinning everything down is that so many people come and go that a lot of the stories either get mixed up or lost entirely in the day-to-day shuffle of such a large facility with a largely transitory population. That the facility is a jail may add another layer of difficulty. Witness credibility is of substantial importance in the information-gathering phase of a paranormal investigation. Jail inmates' credibility may be difficult to establish. Some may have a variety of mental health issues (diagnosed or otherwise) or may be attempting to use a paranormal claim as part of a legal defense strategy.[4] Nevertheless, we have identified several stories of interest.

A claim made by multiple different inmates over the years is that during the night, they've heard the sound of keys jingling. If you think this doesn't sound particularly anomalous in a jail, you'd be correct. In fact, jingling the keys is known within the jail as the sound the sheriffs make to alert the inmates they're doing their rounds. Not much of a ghost story. However, the inmates maintain that they've heard the sounds

4   Though neither a Colorado case nor one worked on directly by the Rocky Mountain Paranormal Research Society, the *Amityville Horror* case is a good example. That book (and later, ever-growing film franchise) was based on a claim made by the Lutz family after moving into the house in which the (real) Ronald DeFeo murders took place. Years later, DeFeo's own defense attorney, William Weber, admitted to inventing the entire paranormal story "over many bottles of wine" with the Lutzes. For the Lutz family, it was about profit. On the part of DeFeo and Weber, it was about complicating the murder trial and coming up with a potentially novel defense strategy.

of keys even when no one is present. Perhaps just strange acoustics carrying the sound from somewhere else? Perhaps the ghost of a former guard? Perhaps something entirely different?

Staff and inmates both report hearing the sound of someone pulling on the cell doors. Again, this is not unexpected. Like the jingling of keys, pulling on doors is a normal part of the officers' rounds. They tug at the doors to make sure they're closed and locked. But they say they've heard the sounds even when no one was walking the rounds.

One morning, an inmate asked a prison supervisor why someone had been placed in his cell the previous evening. This inmate was supposed to be kept alone. When asked what he'd seen, he said during the night he got up to use the toilet and saw another inmate resting on the top bunk. He went to bed, figuring he'd raise the issue with staff the following morning, but when he awoke he was again alone in his cell. Staff members confirmed that he'd been alone the entire night.

Multiple staff members have reported seeing "shadows" moving around on various tiers of the cells throughout the facility. The cell blocks are arranged into three levels. Cells line the walls, with an open space in the center. Catwalks and stairways in that open space provide access to the cells on all three floors.

Figure 10.3. Denver County Jail cell block. Photo: Bryan Bonner.

There is also a "pipe chase" utility access tunnel located between units 6 and 8. Staff members have reported seeing a person or a shadow at the end of the tunnel. In various versions of the story, this individual or entity either vanishes or appears to walk through a wall.

Figure 10.4. Denver County Jail "pipe chase." Photo: Bryan Bonner.

Some inmates refused to use the community shower on the first level of unit 6. Apparently they "felt" something odd there.

Staff members working in or around the infirmary at the rear of unit 2 have reported seeing a woman there who didn't belong. The mysterious woman is said to have vanished in front of multiple witnesses.

One deputy sheriff who was onsite during our investigation told us a story that occurred when he was working in the back hall area that connects buildings 6 and 8. On one occasion, he felt someone push his shoulder. As expected, he spun around to see why a coworker had pushed him, but no one was there.

Several others who we interviewed mentioned they've heard voices at the back of building 6 even when no one was present.

Other staff members reported activity in the Law Library. Books and papers have been moved out of place without explanation. Individuals have seen a person standing in the room at night when the library was supposed to be closed and empty. Of course, when a sheriff sees someone where they're not supposed to be in a jail, they don't just ignore it. Suspecting an inmate may be out of place, they've immediately investigated whenever this mysterious person is seen. But they always found the room to be empty.

Finally, there's a story (unfortunately) directly connected to some of the history we outlined in the previous section. Recall Deputy Osborne, murdered by inmates near the "cage," a secure location for officers behind the bird's nest where jail personnel can monitor the inmates. Several staff members report seeing his ghost in the location of his death. Inmates have reported seeing a guard in an incorrect uniform around the same location. When asked to elaborate, they've described Deputy Osborne's uniform, complexion, hair style, and black square framed glasses.

Our Investigation

Over the years, we've investigated a lot of different places. Some of them are easier than others. Despite the ghost stories, we never really expected to be able to investigate in an active jail, however. We may be willing to find ways to work under a lot of different conditions, but there's simply no way that we can do our work in an active jail. The number of inmates in such a facility would make it impossible for us to properly account for any noises we might hear. The need for security would prohibitively restrict our movements and activities. Equipment we bring on an investigation might pose a safety risk. It's one of very few places we just wouldn't expect to be able to investigate.[5]

As luck would have it, an unexpected opportunity presented itself and we seized it. Because the Denver County Jail was to undergo substantial renovations, many buildings had been evacuated of inmates. We were invited to conduct an investigation in the newly-evacuated buildings in the short time between their evacuation and their eventual demolition. This gave us the opportunity (as far as we know unprecedented in the world of paranormal investigation) to examine an active jail still, for all intents and purposes, in exactly the condition it would have been while inmates were present, only without the inmates present.

A frequent frustration in paranormal investigation is the limited amount of time one has to conduct an investigation. The case of the Denver County Jail was both better and worse in this regard. It was better in the sense that, while we often are given only one (occasionally two) nights at a time to conduct an investigation, the Denver County Jail gave us four nights over two weekends: May 21, 22, 28, and 29 of 2010. On the unfortunate side, the buildings we were working on were slated for demolition, so there would be no second chance to come back at another time for any follow up research. We were given a golden opportunity, but only *one* opportunity, to get this right. No pressure.

While the areas to which we were granted access were empty of staff and inmates, we had to remain cognizant of the fact that the remainder of the campus was still an active jail. When investigating a paranormal claim, it's necessary to understand what's happening not only in the room or building in which one is working, but everywhere else in the area. Recall from Chapter 1 how sounds from a freeway created the auditory illusion of a choir singing. At Denver County Jail, we had to make sure we accounted for the activities of the jail when interpreting any results.

5   And we mean that. We've been to some weird places over the years. Chapter 6 described our work in a cave. The next part of this book will describe some interesting private residences. Future volumes will tell you all about times we've investigated inside an active resort and casino (itself not an easy feat), mines, cemeteries, busy roads, and even a courthouse. There are very few challenges we'd ever turn down.

Another challenge that presented itself was that, despite the buildings' evacuation and despite the four days we had to work, even the empty sections of the campus provided simply too much space for our small team to cover in the time allotted, so we had to make sure we focused on the most important areas for our investigation. Even still, we set up more equipment than usual to make sure we covered as much ground as possible.

One of the most remarkable things about the Denver County Jail investigation was simply the facility itself. As we toured the property and began setting up our equipment, we were able to get a sense of what life was like for the inmates who'd been living there. The tiny cells, small bunk beds, and exposed toilets painted a vivid portrait of a rather dismal life, devoid of privacy and with only the most limited of comforts.

Figure 10.5. A tiny cell at Denver County Jail. Photo: Bryan Bonner.

We found one hiding place in which an inmate had secretly kept a stash not of the kinds of contraband one might expect in a prison, but of simple bars of soap.

Figure 10.6. An inmate's hidden stash of soap. Photo: Bryan Bonner.

Inmates were, perhaps as one might expect, strictly forbidden to write or draw on the walls, even in their own cells. One might also expect that this rule was often ignored. Graffiti was a common sight as we toured the facility. Because the buildings were slated for demolition, we knew that these messages and artworks would soon be lost, so we considered it part of our duty to document as many of them as possible. Though we can't begin to share all the photos we took in the pages of this book, we wanted to share a few representative samples.

One wall contained numerous messages, expressions of the inmates' anger, despair, loneliness, and a host of other emotions.

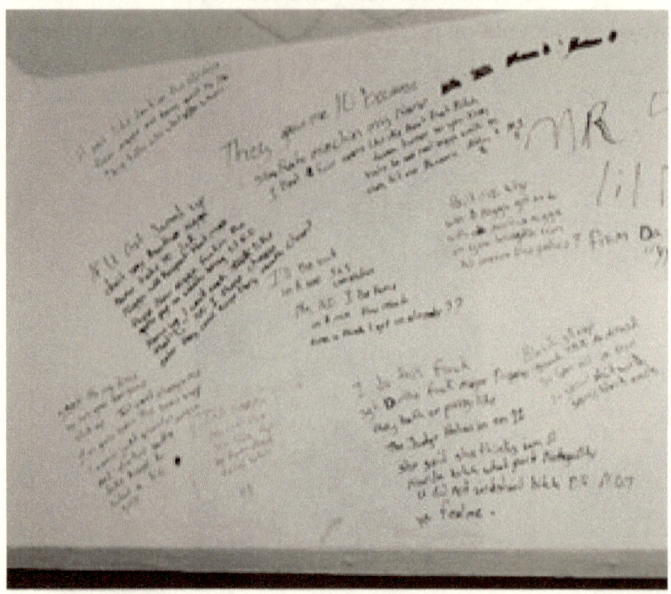

Figure 10.7. Graffiti written by Denver County Jail inmates. Photo: Bryan Bonner.

But the one that really caught our attention was a particular cell in which an inmate had decorated the walls with truly astonishing works of art. Because creation even of these attractive artworks was a prohibited activity, we were told the entire scene had been created by a single inmate working in a single night. The scope and quality of the artworks makes us a little bit skeptical. It's difficult to imagine how a person could create so much in so little time. On the other hand, we're continually amazed by the human capacity to achieve seemingly impossible things. It's not beyond the scope of possibility to think that an individual with a particular talent and nothing else to occupy his time might have actually pulled off such a feat. Either way, we're proud to be able to show you something that would otherwise have been completely lost to history. We were unable to determine the inmate's identity, so unfortunately we cannot give proper credit.

Figure 10.8. An inmate's artwork. Artist unknown. Photo: Bryan Bonner.

We arrived the first night, May 21, 2010, at approximately 7:00 p.m. and following a quick tour of the facility began setting up our equipment in unit 8. We established a base of operations in the cage behind the bird's nest where staff could monitor the inmates' movements and activities. Additionally, we set up remote monitoring equipment in the following locations:

- One video camera and one microphone in the pipe chase,
- One video camera near the entrance, with a view of the entire length of the main hall,
- One video camera on the third tier also monitoring the entire room,
- One video camera in cell 6 (the location of the suicide by hanging), with a view of the bed and cell door, as well as a video camera outside the cell looking in,
- One "natural" (DC) EMF meter, one thermometer, and one seismometer in cell 6, within view of the camera,
- One microphone just outside cell 6,
- One video camera located opposite the cells on the first tier, pointed toward the wall of cells near the front,
- One additional video camera opposite the wall of cells, monitoring the cells toward the rear,
- One thermometer at the main entrance, and
- One stand-mounted microphone on the second tier near the main entrance.

Additionally, we had the requisite mixing boards, monitors, and computers at the base camp to receive, monitor, and record all of the incoming data. We finished setting up around 10:30 p.m. and conducted a sweep of the facility with both AC and DC EMF meters to establish baselines. Throughout the remainder of the evening, we conducted hourly sweeps with these devices.

We noted right from the start that the cells had unusually high levels (3-4 milligauss) of man-made (AC) EMF located near the back wall. This was due to the main wiring and utility tunnel located behind that wall. Readings on the AC EMF meter were consistent throughout the evening. Readings of natural (DC) EMF were below measurable levels and consistent throughout the evening. It's important to note that there is evidence that high levels of electromagnetic radiation can, in some individuals, under some circumstances, result in hallucinations or religious experiences. The levels we measured in the Jail, while higher than in most buildings, were almost certainly insufficient to cause any such phenomenon.

Temperature readings throughout the evening were also consistent, with slight variations of 2-3 degrees Fahrenheit due to changing exterior air temperature.

We pause our story to offer a word of advice should you ever find yourself in a position to conduct a paranormal investigation in a jail. Among the equipment we usually carry with us are UV lights and other equipment that can be used to detect traces of blood or other bodily fluids. Because ghost stories often either involve blood themselves or take place at the sites of violence, that can be an important piece of the puzzle. However, if you ever turn on a UV light in a jail, you'll quickly realize that the walls light up like a carnival hall, and that's not the kind of sight you're quick to forget. So maybe leave the UV in the trunk on such an investigation. We turned it on once, and that was more than enough for us.

Things got a little more interesting when we heard what sounded like cell doors being pulled against their locks. This was a sort of loud thumping sound. Initially, we couldn't identify which cell doors were being manipulated and the ghost story related to some phantom guard pulling on the cell doors even when no guards were walking their rounds immediately came to mind. One of our team members even thought he heard what sounded like footsteps following one of these loud thumps.

Additionally, we heard a very low humming kind of sound on the first tier at the far end of the main room. This hum was most noticeable in cell 12. When one of our team members stepped into that cell, he immediately had a dreadful yet difficult to describe feeling that something was wrong in the cell.

For the first night, we were unable to determine the sources of these sounds or feelings. Unlike most ghost hunting groups, we did not immediately chalk them up to a ghost. And unlike most skeptical groups, we did not immediately assume a mundane explanation. Instead, we noted the anomalies and planned to investigate them further on the following nights. Before we left, we had the opportunity to interview four staff members, who all confirmed that they'd witnessed some of the alleged paranormal happenings within the jail.

A brief note regarding witness credibility: there's a temptation to immediately assume that sheriffs and other staff members are more credible sources than inmates. To an extent, this may be true, at least under some circumstances. It's also true that many law enforcement officers are trained to notice details that others might miss. This is a useful skill in solving or stopping crime, of course, but also may be useful in the case of paranormal investigation. However, it's also possible that these powers of observation could be overactive in the paranormal setting. Something we've noticed over the years is that police and other first responders often have a superstitious streak. We suspect this is partly because of the difficult job they have to do, which also accounts for the "gallows humor" common among these professions. It may also partly be that precisely because these individuals are so well trained to notice patterns or anomalies, they may be hypersensitive to them and take special note of perfectly mundane things that others might not even detect. Therefore, we adopt a philosophy of taking everyone at their word with regard to what they *think* they saw (unless they give us good reason to suspect a hoax, which is rare), but also to always maintain healthy skepticism regarding what they *actually* saw, until we've conducted further investigation.

Our second night of monitoring began much as the first did. We arrived at about 8:00 p.m. and set up as before. However, we made a few changes to our monitoring equipment in light of events from the previous evening. Specifically, we moved the camera from the far end of the first tier closer to cell 12, which had become of considerable interest to us. We began our quiet monitoring and hourly EMF sweeps at around 10:00 p.m. Readings continued to remain steady throughout the evening.

Because cell 12 had become of particular interest to us, we spent some time looking into it in more detail. We used a seismometer and sound pressure meter in an attempt to determine the source of the humming sound. It paid off. We were able to determine that the humming sound was coming from a combination of the main fans circulating air for the facility, and the fans behind the community showers located in units 6 and 8. Experiments with turning the shower fans on and off also yielded interesting details.

With the shower fans off, the ambient sound level was consistently around 62 decibels (db). With them on, the level climbed to as high as 72 db. For comparison, human speech is usually around 60 db, a shower or dishwasher around 70 db, and a blender usually around 80 db. That is to say, while this is not a level of sound that is immediately threatening to human hearing, it is louder than normal background noise. Why, then, did nobody notice it before? It turns out that the bulk of the sound was in a particularly low-frequency range. With the shower fans on, it became louder and slightly higher in pitch; with the fans off, it was lower in both volume and pitch. Specifically, we noticed that many of the tones in the "hum" we heard were actually in the

10-18 Hz range, below the range of human perception.

This proved to be an important discovery not only because it provided an explanation for the hum we'd heard but because it provided a more than plausible explanation for some of the paranormal reports. It's now well documented that infrasound (that is, noise below human hearing) at around 18-19 Hz (often reported specifically as 18.9 Hz) can cause low-grade hallucinations and feelings of dread. The resonant frequency of the human eye is within this range, so loud infrasound noise may cause changes in visual perception or hallucinations of strange shapes, shadows, or movements. Additionally, for reasons not fully understood (which may be related to these hallucinations or may be a product of other brain structures), frequencies in the same range cause feelings of profound unease.[6]

Recall from earlier that inmates didn't like to use these community showers because they felt like "something" was in there with them. We were also told that inmates in cell 12 had reported strange feelings, like someone was in the cell with them. Likely these phenomena were caused by the infrasound. We experienced it ourselves. Inside the cell with the fans on, our people felt extremely uncomfortable. Outside the cell and/or with the fans off, it seemed like just another cell.

In fact, this discovery has made this one of our favorite investigations. We'd read over the years that infrasound could be responsible for some paranormal claims. To have the opportunity to witness and document the phenomenon in person was quite exciting.

That still left us with the question of the thumping noises that sounded like someone tugging on the cell doors. To investigate this, we started by stationing three members of our team in the main room on the first tier to watch the cell doors and see if we could identify which, if any, were opening or closing the previous evening. The sounds persisted during this time. After about 45 minutes of observation, one of the team members reported seeing door 2 moving slightly at the time the sound was heard. We then moved a different team member to watch that door directly. About 15 minutes later, this individual reported the same thing. We hadn't solved the mystery, but at least we thought we were on the right track.

We tried placing a seismometer near the door to see if we could detect any vibrations that might either cause the door to move or that might result from the door moving. Additionally, we placed a video camera to observe the door directly. The occurrence seemed to be happening approximately every 15-20 minutes, so we thought

6   Tandy, V. & Lawrence, T. R. (1998). The ghost in the machine. *Journal – Society for Psychical Research, 62*: 360-364.

Tandy, V., & Street, P. (2000). Something in the cellar. *Journal – Society for Psychical Research, 64:* 129-140.

Landstrom, U., & Pelmar, P. L. (1993). Infrasound: A Short Review. *Journal of Low Frequency Noise and Vibration, 12*(3): 72-74.

there might be some mechanical issue moving the door at that interval.

However, as soon as we got all the monitoring equipment set up, the sound and movements started affecting different doors than the one we were monitoring. We were never able to record the sound or movement and thus the mystery persisted.

We returned the following weekend for additional investigations in Building 6. For this "second" investigation (the third and fourth nights), we changed our setup a bit. We established a base camp on the first floor cage and set up the following monitoring equipment:

- One video camera aimed at the door of cell 1 on each of the three tiers (these cameras were intended to capture any door movements, but were repositioned at 12:00 to look into the main room of building 6),
- One video camera mounted on the cage to monitor the length of the building,
- One video camera mounted on the second tier railing near the first cell, also monitoring the length of the building,
- One video camera on each tier's pipe chase, along with a microphone in the first floor pipe chase,
- One microphone on the second tier near the murder site outside the cage,
- One video camera on the second tier pointed toward the murder site outside the cage,
- An additional video camera pointed toward the second tier murder site as well as up the stairs,
- One video camera on the second tear facing the guard area,
- One video camera pointed down the second tier hallway toward the murder site,
- One thermometer at the murder site near the cage,
- Two control objects (balls) located at the murder site, within view of the cameras,
- One seismometer on the desk in the second tier cage,
- One control object (ball) on the stairs leading from the second to third tier, and
- One thermometer at the base camp in the first tier cage.

Initially, we assigned team members to monitor several locations: just outside the guard area on the first tier, at the top of the stairs on the third tier, sitting at a table at the end of the building, at the murder location on the second tier near the cage.

Things started off quiet enough. EMF, temperature, and seismometer readings weren't revealing anything out of the ordinary. Manmade electromagnetic fields continued to be above normal levels, but we determined (as we did on the previous weekend) that this was due to the building's wiring and utilities. We did observe readings as high as 50 milligauss in the "hottest" spots. This is comparable to readings one might get while standing near electrical appliances in the home.

Figure 10.9. "Control objects" at the site of Deputy Osborne's murder. Photo: Bryan Bonner.

Around 2:00 a.m., though, one of our observers noticed what appeared to be a shadow around cell 5 on the first tier. At the same time, he noticed that the "Tx/Rx" lights on his two-way radio had begun to flash. "Tx" stands for "radio transmitter" and "Rx" stands for "radio receiver." These lights should activate when data is being sent or received by the radio. We added a camera to monitor the radio and cell 5. Additionally, we moved the team member from the top of the stairs to this location so there'd be multiple witnesses if anything else happened. Both of them reported seeing a similar shadow in cell 5, but nothing was recorded on video.[7] We weren't able to record or determine anything further.

None of the control objects were seen to move on any of the video recordings. Additionally, before and after photographs of their locations revealed they remained in place the entire time.

One humorous incident occurred when one of our team members needed to rest for a while (a common occurrence on these late-night investigations). We directed her

---

7   A common claim by some ghost hunting groups is that events or entities show up in their photographs that were not seen by the people present when the recordings were taken. In our experience, the opposite is far more frequently the case. We tend to view anomalies in photographs that were not witnessed in person as likely due to photographer error. However, when our team members see something in person that is not recorded on video, this is much more common, much more mysterious, and much more frustrating.

to a particular room to take a nap. We didn't tell her until later that she was napping just feet away from where a murder took place. But that's just the way things go when you're on an investigation with Rocky Mountain Paranormal.

Overall, we were quite grateful for the opportunity to explore this facility. It's something not a lot of people get to see unless they're incarcerated or work in law enforcement—and even then, they don't get to see it under the conditions we did. And we did experience some anomalous phenomena during our investigation, some of which we were able to solve and some of which remain mysterious to us.

We suspect that a lot of the ghost stories may have grown from relatively mild phenomena like those we witnessed. If someone sees something unusual and tells someone else, who tells someone else, there's always the chance (even the likelihood) that the story will grow over time. Almost by definition, that's how urban legends work. In the case of Denver County Jail, the inmates provide a literally captive audience and an isolated social network, where these stories can pass around from inmate to inmate for months or even years, much like a game of telephone.

In the end, we were left with plenty of questions and only a few answers.

References and Further Reading

Field, A. J. (2005). *Mainliner Denver: The Bombing of Flight 629*. Boulder, CO: Johnson Books.

Landstrom, U., & Pelmar, P. L. (1993). Infrasound: A Short Review. *Journal of Low Frequency Noise and Vibration, 12*(3): 72-74.

Poeppelmeyer, M. (2019). *Finding My Father: Beyond Tragedy, Through Trauma, and Into Freedom. A True Story*. Powell, OH: Author Academy Elite.

Tandy, V. & Lawrence, T. R. (1998). The ghost in the machine. *Journal – Society for Psychical Research, 62*: 360-364.

Tandy, V., & Street, P. (2000). Something in the cellar. *Journal – Society for Psychical Research, 64*: 129-140.

11

The Hatchet Lady and the Grizzled Man: Red Rocks Trading Post

Red Rocks, so named because of its characteristic large, red-colored sandstone structures, is a mountain park operated by the City of Denver. It's probably best known for Red Rocks Amphitheatre, an open-air music venue built directly into one of those rock structures within the park's boundaries. With a capacity of nearly ten thousand people, it's one of the largest music and performing arts venues in the State of Colorado.

But our story is not about the Amphitheatre and is only partly related to the park. Our story is about the Red Rocks Trading Post, a historic building reminiscent of a gold rush era trading post, which now maintains a similar function as a gift shop for park visitors. It's also allegedly the home of some restless spirits, which caught our attention.

Red Rocks Trading Post is located at 17900 Trading Post Road in Morrison, Colorado, and is frequently used as a first stop on a tour of Red Rocks Park.

Figure 11.1. Red Rocks Trading Post: Photo: Bryan Bonner.

The History

As we've done with some of our other geologically interesting investigation sites, we will present both the natural and human history. We do this because when one visits Red Rocks Park, probably the first thing that crosses one's mind is a thought regarding the natural beauty of the area. Second, likely, one might wonder how these magnificent natural structures came to be there.

You may recall either from an earlier chapter in this book or from a geology class you once took in school (or perhaps a visit to the magnificent Denver Museum of Nature and Science) that Colorado was not always a landlocked state located roughly a mile above sea level. In the geological past, our region was part of a vast sea (which is why one can still find fossils of ancient sea-dwelling creatures in the Rocky Mountains).

About 300 million years ago, Colorado consisted of a landmass surrounded by these prehistoric oceans. The rising and falling of sea levels and land masses over time resulted in sandy beaches that eventually formed a massive sandstone formation called the Fountain Formation. The deposits that formed this sandstone originally eroded from the early Rocky Mountains.

Red colors in the Red Rocks stones comes from a substantial iron content in the deposits that formed the sandstone. When iron oxidizes, it turns a reddish color. You would recognize this process in more familiar settings as rust. When oxygen bonds with iron, it forms these red-colored iron oxides.

The original Rocky Mountains formed some 65 million years ago but had eroded away by about 40 million years ago. Their material washed eastward forming the Fountain Formation from which the Red Rocks Park emerged.

Recall from Chapter 6 that the current Rocky Mountains were lifted by plate tectonic forces. This uplifting included the Fountain Formation. As the mountains emerged, they lifted and split the Fountain Formation, turning the sandstone on its edge. This is why many of the most famous rock structures seem to protrude at steep and unusual angles from the ground.

The story doesn't end there, though. Lifting these massive sandstone monoliths undoubtedly would have resulted in a spectacular sight on its own, but other geological processes have intervened in the last 40 million years or so. Because the Fountain Formation consists of deposits of many different types of material, it erodes unevenly. As wind and water erosion continue to sculpt the structures, the emerging work of natural art includes stripes, pits, crevices, caves, and visible layers of different colors.

The human history of the area, as far as we can tell, begins with the Ute, who were known to frequent the area. One Ute Chief named Colorow (c. 1810 – 1888) was known to camp in what is now called Colorow's Cave just south of Red Rocks Park. Colorow was a famous warrior and participant in many of the Ute's treaty negotiations with the U.S. government. He fought in the Meeker Massacre[1] in 1879 and was attacked along with his family in 1887 in an event that's come to be known as Colorow's War.

American history in the region began when Whitman Cross of the U.S. Geological Survey described and named the Fountain Formation in 1849.

Residents of the area now known as Morrison named the area the "Garden of the Angels." A parade in honor of the name's dedication was held on July 4, 1870. The parade was led by a local judge and farmer named Martin Van Buren Luther[2], who'd suggested the name.

In the early 1900s, much of the land was owned by John Brisben Walker (1847-1931), a publisher and entrepreneur who brought *Cosmopolitan Magazine* to massive success before selling it to Wllliam Randolph Hearst in 1905. Recognizing the natural acoustics of what would eventually become Red Rocks Amphitheatre, he had a vision of the region becoming a park and cultural institution. He erected a temporary stage and held some musical performances as early as 1909. In addition, with his son John Jr., he held baseball games and races in the park. During this time, the park was also known as the "Garden of the Titans," a move away from Judge Luther's original name.

In the 1920s, a U.S. Army expedition under the command of Stephen H. Long passed through the region. A member of the company, one Edwin James, remarked that Red Rocks had a "striking resemblance to colossal ruins." His statement was somewhat prescient—more than a century later, the Red Rocks Amphitheatre would open in 1941, albeit with a much lower capacity than the Colosseum, which may have held as many as 80,000 spectators at its peak.

Eventually, following years of negotiations, Walker sold the land to the City of Denver (then under the supervision of Mayor Benjamin Stapleton) for $54,133 (equivalent to just over $900,000 in 2022). This allowed for the creation of a city park, which officially changed its name to "Red Rocks" in 1928. Construction on the Amphitheatre, designed by architect Burnham Hoyt[3], began in 1936 and was completed in 1941, though that's where we'll leave the Amphitheatre portion of the story.

The Trading Post was constructed between 1930 and 1931 following designs by Denver-based architect W. R. Rosche in the Pueblo Revival style. Its design was intended to complement the surrounding natural "architecture" in the park. We think it was fairly successful in that goal. Upon its opening, it was first called the Indian Concession House and later the Indian Trading Post. In addition to its function as a trading

---

1   The site of the Meeker Massacre, in what is now Rio Blanco County in northwestern Colorado, is also reputed to be haunted. Rocky Mountain Paranormal has not (yet) had the opportunity to investigate that area.

2   Not the eighth President of the United States. President Martin Van Buren died at his home in Kinderhook, New York in 1862.

3   Also responsible for dozens of Denver landmarks, including the Denver Press Club—see Chapter 4.

post, it was originally home to a small museum focused on local wildlife and geology.

When it was first opened, it was also home to a caretaker who held a residence in the lower level of the building. This residence remained occupied until 1983, when it was converted to office and storage space.

In 1994, Red Rocks Trading Post was recognized as a Denver Landmark. It was recognized along with the Park and Amphitheatre as a National Historic Landmark in 2016.

Paranormal Claims

Paranormal claims can be divided productively into the general and the specific. General claims include things like items moving around, lights turning on or off, cold spots, and so forth. People may attribute these phenomena to ghosts, but don't necessarily attribute them to a specific identifiable ghost. Specific claims, on the other hand, are focused around a particular ghostly or paranormal figure or entity. Of course there's some gray area between the two types of claims (for instance, the claim may be of an object moving, but it may be attributed to a specific ghost), but it's useful to distinguish between the two anyway. Specific claims are more susceptible to historical analysis because a researcher can attempt to determine relevant facts regarding the alleged ghost's identity. General claims are typically subject only to techniques related to direct observation.

The Red Rocks Trading Post has its share of both classes of claims related to ghostly activity, as well as some non-ghost paranormal claims we find rather interesting.

The first and probably most famous specific entity claimed to haunt the Trading Post has become known as the "Hatchet Lady." Multiple variations of the story exist, which we'll discuss in turn, but the common thread between all of them is that this individual is a female spirit who visibly manifests and carries a hatchet.

Fans of horror movies might not think twice about this sort of thing. Hatchet-wielding spirits seem to fit well in the genre. We're accustomed to violent ghosts in works of fiction. But in reality, the Hatchet Lady is a fairly unusual paranormal claim. Most paranormal claims involve less direct manifestations. Spirits that visibly appear to witnesses are not unheard of, but they're the minority of paranormal claims. Furthermore, most claims of ghostly activity are not as threatening or frightening as a spirit carrying a hatchet. Again, such things are not unheard of, but if you're read the previous chapters in this book, you'll have noticed that the vast majority of ghosts are not claimed to directly threaten the living.[4]

4   One thing to keep in mind, particularly if you're frightened because you think your house is haunted, is that there are no documented or verifiable claims of any ghost or demon ever causing physical harm to a living person. The few stories along those lines that exist are both in the extreme minority and have not been properly verified.

The Hatchet Lady, nevertheless, is said to roam the Trading Post and the surrounding areas of Red Rocks Park. Encounters with or stories about this spirit seem to fall into three groups, each providing slightly different details on the story.

1) A former concessions stand worker at the Park and Trading Post said the spirit might be that of one old Mrs. Johnson of the Johnson Clan. According to this witness, Mrs. Johnson used to ride her horse around the property to stop young couples from using the region as a "courting area." She used to pull her jacket up over her head to appear headless and carried a hatchet to frighten the young lovers away.

2) Others have claimed that the Hatchet Hady is rather the spirit of a homeless woman who lived in the Red Rocks canyon during the 1950s. In this version of the story, her spirit still wanders the area carrying her hatchet and chasing away visitors.

3) Finally, there's a version of the story in which the Hatchet Lady is a naked woman carrying a hatchet. In this most horrifying version, she doesn't just scare away visitors, but is said to kill children who come too close to the cave in which she used to live. After committing these murders, she reportedly hides the bodies and severed appendages under rocks in the area.

Typically, despite our overall skeptical approach to paranormal investigation, we at least *hope* that some of the ghost stories turn out to be true. How amazing would it be not only to think something like that actually exists, but to be the group responsible for discovering or documenting it? However, in the case of at least the third version of the Hatchet Lady story, we can honestly say we really hope it isn't true. That's not going to stop us from looking into the claims, though.

A second specific entity said to haunt the Trading Post and surrounding area is described as a "grizzled looking old man," and so we will refer to him as the Grizzled Man for the purposes of our discussion. Reports have him at approximately five feet and five inches in height, with dirty brown hair and a beard, often carrying a bottle in his hand. Encounters with this spirit follow a predictable pattern: he appears to the witness briefly and then vanishes. Some have speculated that he is the spirit of a miner who used to hunt for treasure in the region. Despite his unkempt appearance, the Grizzled Man seems to present a substantially less threatening figure than the Hatchet Lady.

Additional reports of apparitions tend to be of Native American spirits. These range widely in their details. Some people report seeing a single individual. Others have reported witnessing entire ceremonial dances performed by the ghostly apparitions. No specific identities seem to be ascribed to any of these spirits in the various stories we've found over the years.

More recent reports of activity at the Trading Post are more of the generic quality.

Items are reported missing or are found where they're not supposed to be. Staff members have reported a coffee mug with the name "Randy" written on it falling off the same shelf every day. Other employees have reported the sound of a dog scratching at all of the exterior doors, but finding no dog present when they've investigated. One individual claimed that several years ago, around 2:00 a.m. while working a late shift at the Trading Post, he heard "loud boot steps" approaching and then felt a chill as something brushed his back. No one else was present.

One phenomenon reported by several of the gift shop employees centers on a door that leads to the rear storage area. According to their claim, the door will sometimes lock itself without any human intervention. One of the staff members reports actually seeing the lock moving on its own. After this event, staff members could no longer open the door. Maintenance workers had to break the door handle to get the door open, after which the handle was replaced. To our knowledge, that put a stop to the self-locking door phenomenon.

In 2005, one employee reported seeing the large door on the cold storage room open by itself. Both the handle and the door moved, yet no one else was around.

Because the Red Rocks Trading Post is an active business, sometimes items get moved around not by a paranormal force but simply as a matter of renovation. In a prior configuration of the shop, a set of cock wind chimes were located in the main room near the back by the fireplace. A staff member told us these would sometimes move and chime on their own when there was no draft or breeze.

Figure 11.2. Red Rocks Trading Post basement stairs. Photo: Bryan Bonner.

Several people have told us that they've heard the sound of footsteps on the stairs leading to the office on the second floor. Inevitably, the footsteps stop partway, never reaching the top of the stairs. Footsteps have also been reported on the ramp in front of the candy counter.

Speaking of hearing strange things, one employee reported hearing their name called while working in the laundry room in the basement. As is usually the case with these kinds of stories, no one else was present in the basement at the time.

One enterprising staff member decided to do a bit of paranormal investigation of his own and try an experiment. He left an audio recorder in the basement one evening after everyone left for the night. Upon playing back the audio recording later, he heard sounds of breathing and boxes moving in the empty room. Unfortunately, we haven't managed to get our hands on this audio for a more detailed analysis.

Figure 11.3. Basement storage area at Red Rocks Trading Post. Photo: Bryan Bonner.

In an adjacent basement room to the one in which the audio recording was taken, another employee was once stocking some shelves. While she was working, she says, she saw a box coming right at her from across the room.

Most of these claims from the employees have occurred between around dusk and approximately 2:00 a.m. when the employees are picking up supplies and setting up for the following day.

Indeed, most (but not all) paranormal events seem to happen at night. Whether that's because those are the hours when paranormal entities are most active, due to humans' natural fear of the dark, or because nighttime is typically the time when things are quiet and we are most likely to hear things without being overwhelmed by the sounds of the hustle and bustle of the daylight hours remains a matter of speculation.

There's one additional story that caught our attention. One day, the staff at the Trading Post received a parcel in the mail containing an unusual item: a rock from the Red Rocks Park. An enclosed letter explained that the sender had taken the rock from the park as a souvenir, not realizing it is considered back luck to do so. The letter went on to explain that the sender had experienced an unusual run of bad luck in the weeks since taking the rock and hoped that they might break the curse by sending it back to its source.

Our Investigation

Our work at the Red Rocks Trading Post began when we were contacted by Denver's local Channel 7 news program about the reports of paranormal activity at the site. Because the on-site investigation was to be documented by the television crew for their report, our time at the venue and our activities were somewhat limited, but we nevertheless did our best to proceed with the investigation as we normally would.

Before we discuss our on-site investigation, however, we should tie up a few loose ends regarding background and historic research because those have some bearing on the paranormal claims.

With regard to the Hatchet Lady, we wanted to see if we could determine any potential historic figures that might be associated with the entity. Unfortunately, two of the three variations of the story don't have any names attached to them. Indeed, if there were ever a homeless woman living in the Park, there would likely be no records of her identity or activities, so there's really not much we can do.

We were able to determine that there have been several Johnson families who've lived in the region throughout history (unsurprisingly). Because of the reports that the Hatchet Lady may be "Old Mrs. Johnson," we wanted to know if we could pin down a specific Mrs. Johnson to the claim. To date, we've been unable to do so. Johnson is the second most common surname in the United States, so without knowing a first name or which "Johnson clan" is implicated in the story, we hit a dead end.

The third variant of the Hatchet Lady story is of course the most gruesome. It's also another one where it would be all but impossible to track down the identity of the naked hatchet-wielding woman in question because no details of her identity are a part of the lore. However, we have searched newspaper records and found no indication that anyone—ghost or mortal—has been dismembering children and hiding their bodies under rocks in the area. Of course, just because no such bodies (or parts thereof) have been discovered, that doesn't *entirely* discredit the story, but we are of the opinion that if the story were true, there would at least be some evidence of the deaths in the public record. Fortunately, no such records exist.

The story that the Hatchet Lady attacks children is reminiscent of many scary stories parents have told their children over the years to "scare them straight." One of the psychological functions of a childhood belief in any kind of monster or boogeyman seems to be to encourage children to stick to the straight and narrow path because any deviation might result in one's destruction. Indeed, the same message can be found, in a more adult format, throughout the entire horror genre. Don't trespass or whatever lurks behind locked doors might get you. It's even a much-mocked trope that having sex in a horror movie damns one to die at the hands of the slasher. Likely that became a cliché in horror because the message resonated with audiences—casual sex carries all kinds of risks. The horror story's job is to exaggerate those risks and provide its warning in a narratively engaging manner.

Our opinion (and it is just an opinion, as there's no particular evidence to support this or any other interpretation) is that the third and most gruesome version of the Hatchet Lady story is the result of the merging of other Hatchet Lady stories with the various cautionary tales parents have been telling their children since time immemorial.

We were also interested in the story of the rock being mailed back to the Trading Post to break a curse. Stories like that have come across our desk several times over the years, and our members even keep several such "cursed" items in our own cabinets of curiosities. But we were unaware of the idea that taking a rock from Red Rocks Park specifically was supposed to bring bad luck.

During our on-site investigation, we asked the Trading Post staff if they'd let us take the rock in question with us to see if our luck changed for the worse. They didn't go for that idea. The thought occurred to us to take some random rock from the Park, but our ethics prevent us from doing so. It may seem like an inconsequential thing to take a mere rock, but if everyone did so, it would diminish the natural beauty of our public parks and so we will never take anything from any investigation site without permission.

As we were researching the story, though, we found a more recent case in another Colorado location that tells almost exactly the same story. According to a 2020 report in the *Idaho Statesman*, an unidentified individual sent a rock to the Southeast Region of Colorado Parks and Wildlife accompanied by a letter which read: "Someone brought this home to me three years ago. Bad things have been happening ever since. Sure one of you can find which park it belongs in. Thank you, S——." Though the letter was signed, the author's identity has been concealed in press reporting.[5]

A similar story is better-known at Petrified Forest National Park in Arizona. Reports of curses following those who've stolen petrified wood from the park go all the way back to the 1930s and continue to this day. Park rangers still occasionally receive these mailings of petrified wood people have stolen and returned in the hope of breaking a curse.[6]

Whether one chooses to believe in curses or not, most experts seem to think (and we agree) that the folklore surrounding these ideas began in the early 20th century, probably with park rangers and tour guides trying to find a way to scare people out of taking their little souvenirs out of geologically or historically important parks and public spaces.

Finally, before we set out to investigate in person, we wanted to know a little more about the region's geology. Specifically we wanted to know if there were any seismic

5   Capron, M. (2020). Stolen rock is mailed back to Colorado park after recipient has bout of bad luck. *Idaho Statesman,* July 22, 2020.

6   Alexander, K. (2022). The Curse of the Petrified Forest. *Legends of America.* <https://www.legendsofamerica.com/az-petrifiedcurse/>

features (such as low frequency vibrations as described in the previous chapter) that might account for any paranormal claims. A deep dive into this geology makes for fascinating reading for science nerds but revealed no features that might be of interest in investigating paranormal claims.

On the day of the investigation, we arrived at about 5:30 p.m. to conduct interviews and begin setting up our equipment. We chose to establish our base camp at the back of the building near the coffee bar. As is our usual practice, we set up extensive monitoring equipment throughout the site:

- Two camcorders pointing toward each other in the main room with a view of the location where wind chimes were said to move of their own accord,
- One microphone in the main room,
- Two video cameras and one microphone in the back room (where footsteps had been reported),
- One video camera in the basement manager's office,
- One video camera in the basement storage area (where breathing had been heard and boxes were said to move),
- One video camera in the basement laundry room, and
- One "natural" (DC) EMF meter in the basement rear storage area.

Additionally, we used a standard (AC) EMF meter, two thermometers, and 2 seismometers at various locations during hourly readings throughout the investigation.

During initial EMF readings to establish baselines, we discovered a few locations had above-normal electromagnetic fields: the area near the cash register (caused by the register and other computer equipment), the main room near a stuffed buffalo (caused by a powerline running through the wall), the kitchen and coffee bar (caused by the appliances), the crawlspace with the heater (caused by the heater's electrical components), and the rear basement storage area near the main breaker panel (the highest of all the readings, but caused by the breaker panel itself).

Readings remained stable throughout the evening and showed no variance outside of normal fluctuations seen in any location. Temperatures likewise remained stable except for a slight gradual decline caused by decreasing exterior air temperatures.

Unfortunately, that's about all we have to report from this particular investigation. None of our team members witnessed anything out of the ordinary while we were there. When we reviewed the video and audio recordings, we likewise found nothing anomalous or particularly interesting.

That's one of the important lessons of paranormal investigation. Sometimes you get a really interesting case where all kinds of strange things seem to be happening. Other times, you spend the evening in a wonderful venue, but nothing at all happens. Such is the ebb and flow of paranormal research. It's also a way to gauge the credibility of a paranormal investigator. If they have never been on an investigation in which nothing interesting happened, they're probably just playing at ghost hunting because it's fun to scare oneself. On the other hand, if they've never been on an investigation in which something weird did happen (whether or not they were ultimately able to explain it naturally), they're probably not looking hard enough.

Our visit was on the short side, and the reported paranormal phenomena have not occurred at specific intervals, so it's possible either that there's nothing paranormal there, or that the phenomena simply didn't happen to manifest during our short window of observation. The only way to find out which is the correct conclusion would be to conduct further research.

12
Rocky Mountain Paranormal on *Ghost Hunters:* The Elkhorn Lodge

Though not as famous as its near neighbor The Stanley Hotel (see the following two chapters), there's an even older hotel in Estes Park, Colorado with a haunted history of its own. The Elkhorn Lodge, located at 600 Elkhorn Avenue in Estes Park is not only a hotel but a guest ranch, boasting a fascinating history and a substantial rustic campus. It's truly an amazing place for any history buff, and home to some wonderful spooky stories.

When the *Ghost Hunters* television show, now one of the longest-running and most popular paranormal shows in the world, was still fairly new, Rocky Mountain Paranormal alerted the show's hosts to the Elkhorn's ghost stories and some of our members even appeared on the second bonus episode of the second season during the on-screen investigation of the property.

The magnificence of the venue itself coupled with our inside stories about the production of a paranormal television show makes this one of our favorite case files.

Figure 12.1. Elkhorn Lodge. Photo: Bryan Bonner.

The History

The history of the Elkhorn Lodge is inextricably tied with the history of Estes Park itself. As the oldest continuously operating (at least until quite recently) hotel in the Rocky Mountain region, the Elkhorn lodge saw almost the entirety of the growth of Estes Park.

Prior to American settlement, the region now known as Estes Park was dominated by the Ute tribe until the late 1700s or early 1800s, but the history of the township itself really begins with a man named Joel Estes. Having struck it rich in California, Estes relocated to Colorado as part of the gold rush that dominated so much of our state's early history. Born in 1806 in Kentucky, Estes moved around the country quite a bit during his young adult life. After making his fortune in California's gold mines, he moved his family to Colorado to try their hand at farming in Adams County before becoming the first white people to settle in what is now Aurora. But Estes wasn't the kind of man who liked to stay put for too long a time. In 1859, while on a hunting trip, he discovered what is now called the Estes Valley and immediately decided to settle there with his family.

For health reasons, Estes would eventually move to Texas, then back to Colorado, though he didn't return to Estes Park. He died in 1882.

The town was named Estes Park by William Byers, founder of the *Rocky Mountain News*, who was struck by the region's beauty during a visit and saw fit to name it after its first settler. But Estes and Byers were far from the only men to set their sights on the area. One important figure in the town's early history was the Honorable Windham Thomas Wyndham-Quin, the fourth Earl of Dunraven (commonly known simply as Lord Dunraven), an Irish nobleman whose history will feature more prominently in Chapter 14. He visited the region in 1872 and, similarly falling in love with the area, set out to acquire much of the land, an endeavor which we'll discover later eventually became the stuff of local legend.

Just a year or two later, in either late 1873 or early 1874, William E. James visited Colorado from his home in Syracuse, New York. Given what we've read so far, it is perhaps little surprise that Mr. James was also struck by the remarkable beauty and opportunity in the Estes Valley after he stumbled across it on a hunting trip. It didn't take long for him to decide to stay put, so he established a homestead claim on a parcel of land in the Estes Valley.[1] After erecting a small dirt-roofed log cabin as a home

---

1   The Homestead Act of 1862 was a government program designed to encourage expansion into America's western territories and essentially offered free land to anyone who could tame it. Any adult who could establish residency on and improve government surveyed property could, after a period of years, obtain title to the land. There were many goals. One was to reduce poverty. In that particular goal, the act was not all that successful, as those with the resources to tame large parcels of land tended

(which remarkably has survived until the present day), he brought his family—wife Ella McCabe James and three young children: Homer, aged six or seven, Charlie, aged four, and infant Howard—to their new residence.

You'd be forgiven for thinking that was the beginning of the Elkhorn Lodge. It wasn't quite. Winters in Colorado can be harsh, and the cabin was leaky and hardly fit for the raising of the James family. Mr. James was fortuitous enough to arrange to trade properties with the Reverend William McCreery in 1876, allowing the Jameses to settle in their new land along the Fall River on the western edge of what was to become the town of Estes Park.

While raising cattle on his land was somewhat profitable, James quickly realized there was even more profit to be gained through public accommodations. Given that everyone who visited the region seemed to immediately fall in love with it, James reasoned that tourism would be a substantial industry in the future. Perhaps his reasoning was supported by people asking to be allowed to stay on his property when they would come to visit. He began work replacing his family's simple log cabin with a larger resort surrounded by cabins and tents to accommodate lodgers and visitors. The Elkhorn Lodge was thus born, receiving its first guests in 1877.

By 1880, Elkhorn Lodge had grown enough to accommodate as many as forty guests at once, and during the 1890s the dining room could seat as many as two hundred people. It was almost always full for dinner service.

Though the Stanley Hotel we'll discuss shortly may be the more famous Estes Park hotel, the Elkhorn Lodge was the site of much of the town's history. Within its boundaries were Estes Park's first schoolhouse[2], first chapel, first golf course, first icehouse[3], and, indeed, first public hotel.

---

to be those who were not hampered by or destined for poverty in the first place. However, the government also had an interest in growing the national economy by utilizing its newly-acquired lands. Furthermore, the establishment and occupancy of the nation not only on the east coast but across the entire breadth of the continent promised a distinct military advantage: with only two land borders (on the north and south), the United States would be far more secure against foreign invaders than it would be with large swaths of unoccupied and unknown lands in the west.

2   According to Eleanor Hondius, who we'll meet shortly, the school was founded by one Miss Hyde, sister of Albert Alexander Hyde, founder of the Mentholatum ointment company, as a means to educate the James children. After a couple years of this, Mr. James participated in the construction of a proper schoolhouse and establishment of a school district in Estes Park.

3   This building is now called the Woodshed, but its first purpose was housing ice. During the winter months, workers would cut large blocks of ice from the Fall River and store them in the icehouse for preservation and use during the hot summer months. In a time when all we need to do to obtain ice is press a button on our freezer

On January 6, 1880, the James welcomed into their family a new child: Eleanor Estes James, who grew up at the Elkhorn Lodge and whose memoirs provide much of what's known about life at the Lodge.

Unfortunately, young Charles James died of a virus in 1889 at the age of only twenty years. The same year, Ella gave birth to Willie James, but he would not survive past 1890.

Why was the Elkhorn Lodge so successful not only in the beginning, but for so many decades? By all accounts, it owed its success to two primary factors. First of all, as already established, it was quite simply the right location for a guest ranch. It was (and still is, despite modern developments that have turned Estes Park itself into a modern city) the kind of place one would go to commune with nature. That very nature in the region was also what prompted the establishment of the nearby Rocky Mountain National Park by President Woodrow Wilson on January 26, 1915.

The other reason the Lodge may have been so successful was because of the James family themselves. To hear almost anyone tell it, William James was the quintessential gentleman, always ready not only to serve his guests but to regale them with his fishing and hunting tales. The remainder of the family were similarly friendly and welcoming, often letting guests join them for dinner and during their family time gathered around the perpetually roaring fireplace, the only means of heating the building in those days. Such hospitality also extended to the local residents. Those few people who stayed in the unheated buildings during the winters would gather at the Elkhorn Lodge for suppers to keep each other company and pass the long winter nights together, eating whatever the men of the party could successfully hunt earlier in the day.

The same sense of community surrounded the Elkhorn Lodge even as its ownership passed through the generations. At a time some decades later, with Eleanor now in charge of the property, it was still the place to go for social events. And the family in charge would always make sure the events were successful. One woman, identified only as Mrs. X in Eleanor's memoirs, didn't take much interest in the standard events at the Lodge and chose instead to host a flower show. When interest didn't manifest quickly, Eleanor herself took to cajoling locals and visitors into entering the competition to ensure Mrs. X had a successful event.

Mr. James died in 1895, and the management of the Elkhorn Lodge fell to his wife, Ella McCabe James, along with the children Homer, Howard, and Eleanor. The eldest son Homer had previously chosen a career in medicine, having been educated at the University of Colorado, and relocated to New York. Younger son Howard seemed more fit for the hotel life, but in Ella's opinion was too young to manage such a property at the time of William's death, so Homer was summoned back to the Elkhorn to assist. This turned out to be a good turn for the locals who were at that time serviced

door, it's worth reflecting on the amount of work our ancestors put into the things we take for granted.

only by a single physician. As such, Homer James would occasionally step in to offer his medical services. He never billed any of his patients, viewing his profession at that time as managing the hotel and his medical work as simply a kindness to his neighbors.

Dutchman Pieter Hondius, born in 1864 in the Netherlands, immigrated to the United States in 1896 and became a citizen a few years later. For health reasons, he landed in Colorado, and eventually at the Elkhorn Lodge, where all visitors to Estes Park eventually found themselves at the end of "the season." In 1904, his life became tied to the Elkhorn Lodge when he married Eleanor. The Hondius couple would spend the rest of their lives living and working at the Elkhorn Lodge. Meanwhile, Ella eventually died in 1917, handing full control of the Elkhorn Lodge to the James children. Homer would eventually marry a woman named Mary Jane and relocate to California, where he lived until his death in 1958. Howard stayed in Colorado where he married one Edna B. Cobb and contributed to the management of the Elkhorn Lodge until his death in 1928.

After successfully seeing the Elkhorn Lodge through the Great Depression (and raising their son, Pieter Honius, Jr. at the lodge), Eleanor and Pieter eventually sold the property to Howard's estate in the 1940s, though the Hondiuses would remain in the area for the rest of their lives. In her memoirs, Eleanor described a sense of loneliness after the sale because she missed the excitement of tending to the guests for whom she had obvious affection. Nevertheless, though Eleanor had stepped down from the Lodge's management, it's remarkable that such a property remained in the hands of a single family for so long. It's truly a testament to the power of the small American family-owned business. When six of its structures were added to the National Register of Historic Places in 1978, its connection to the James family and their pioneering work in Estes Park were cited in the statement of significance.

However, it wasn't to remain in the family permanently. In 1959 it passed into the hands of a group of Nebraska-based businessmen who later sold it to James Hickman of Tennessee. Eventually it was bought by Jerry and Carol Zahourek in 1990. The Zahoureks began operations in 1991, keeping up with the same traditions and leaving the property (including those structures not granted historic protection) largely intact. However, as the times changed, their labor of love eventually took too much of a financial toll on the Zahoureks and they listed the Elkhorn for sale in 2017.

A company called East Avenue Development purchased the property. In 2021, a substantial renovation project began. As of this writing, the Elkhorn Lodge remains closed as the project continues. Unfortunately some of the old buildings are being lost to make way for newer improvements. However, as far as we can tell, the new owners are cognizant of the property's historic significance and are preserving not only the six structures granted formal protection, but as many of the old buildings as possible, including the old schoolhouse. During the project, some pieces of the buildings that couldn't be saved were sold at auction to fund historic preservation of the remainder of the site. Some of the windows and a sink from the original Elkhorn Lodge now reside permanently in the Rocky Mountain Paranormal collections.

One of the remarkable things about the Elkhorn Lodge is that it was a historic holdout in many ways, providing a traditional refuge from the trials and tribulations of modernity. It wasn't even heated until 2013, meaning it generally operated as a hotel only during the summer months.[4] This was part of the Lodge's appeal, surely, to many people. The Main Lodge consisted of various rooms with accommodations for 65-70 people depending on the particular configurations of the rooms. The rooms themselves were eclectic in their décor, but all of them were rustic and did not feature such amenities/annoyances (depending on one's point of view) as televisions. Electrical outlets were limited. However, it was not completely devoid of modern comforts. All rooms but one had their own bathrooms.

Even the furniture reminds one of a time when things were built to last. When the Lodge was in the process of expanding from a single-family cabin into the multi-building compound it is today, the James family began furnishing the rooms with Stickley furniture. The Stickley furniture company still exists and is much prized by collectors for its high-quality hardwood construction. Collectors and those considering large investments in Stickley furniture were, until the Lodge's recent closure, often sent to the Elkhorn to view its collection of pieces that have stood for over a century.

Hunters in the Estes Park region would often hunt the elk native to the area and then sell the meat to markets in Denver, a highly profitable enterprise that by the early 1900s had severely reduced the elk population. In 1913, the Elkhorn Lodge participated in a project to have 40 specially-built wagons constructed to bring elk back to Estes Park and regenerate the herds. They were successful. Even today, elk watching is a substantial draw for tourists to Estes Park. If you choose to visit Estes Park and see the elk—and we recommend that you do—then for the love of God and all that is holy, *please* heed the rangers' advice and maintain a safe distance from the wildlife. Some of our members, on our own elk-watching trips, have seen far too many people taunting the animals, putting not only their own safety in jeopardy but also that of other tourists and the animals themselves. Elk may look like peaceful animals—and by and large, they are—but they're also wild animals and, yes, a lot bigger and stronger than you think they are.

---

4   Horror fans may recall that the premise of Stephen King's *The Shining* involves a family of winter caretakers for a Colorado mountain hotel during the off season. While that book was famously inspired by a different hotel—see Chapter 14—that particular element remained true of the Elkhorn Lodge much longer than it did of the Stanley Hotel.

Paranormal Claims

In addition to its fascinating history, of course, the Elkhorn Lodge is home to numerous ghost stories. Unfortunately, some of the lore has become a bit contaminated. After *Ghost Hunters* aired their episode about the Elkhorn, paranormal tourists flocked to the location. We're not necessarily complaining. Surely that was good for business and (almost) anything that keeps a historic venue running is okay in our book. However, it's become quite difficult to separate fact from fiction. There's the actual history, the "natural" paranormal lore that developed alongside that history, and then the "artificial" paranormal lore, much of which may have come from shoddy research or mere anecdotes. When these kinds of rumors begin circulating around the Internet, it becomes quite difficult for anyone to determine which stories came from which sources.

Fortunately, much of our own research into the Elkhorn Lodge predates this wave of paranormal tourism, so, while we've kept ourselves informed of more recent developments as well, we think we're in a position to present a higher quality collection of ghost stories than what you might find in a random web search.

Because so much of the Elkhorn's history is connected to the James/Hondius family and to the small community of friends, family, employees, and loyal guests they developed over the decades, the ghost stories center primarily on this cast of historic characters, spanning several decades.

The first such character is James Nugent, better known as Rocky Mountain Jim. He was connected to Lord Dunraven during the early development of Estes Park and acted as a guide for one of Dunraven's friends, known only as Lord H., on a hunting trip. A quarrel ensued and Rocky Mountain Jim feared for his life. Some have suggested Rocky Mountain Jim likely started the quarrel as he was known to have a certain affection for his drink, even to the point that he once drunkenly picked a fight with a grizzly bear and only narrowly escaped with his life, though far from unscathed. However, it's important to note that all of this is the stuff of legend, and not of verifiable history. As reported in Eleanor Hondius's memoirs (and told to her by Abner Sprague, another local), one Griff Evans was hired to kill Rocky Mountain Jim and did indeed shoot the guide. However, the wound didn't kill him immediately. Evans was charged but found not guilty, and Rocky Mountain Jim died from his wound shortly after the trial.

Several people have reported seeing a ghostly apparition of what appears to be a mountain man wandering the Elkhorn property. While this ghost is not conclusively tied to Rocky Mountain Jim by any kind of tangible evidence, most ghost hunters and witnesses believe they've encountered the spirit of James Nugent, residing in death in the same Rocky Mountains he so loved in life.

Another character to feature prominently in the ghostly lore surrounding the Elkhorn Lodge is "Uncle John," as Eleanor called him. His full name was John Henry Size (1847-1941) and though it's not clear how he was related (if at all) to the James family, it's clear they held him in great esteem. He would take residence at the Elkhorn during the summers and became known as "Mr. Fixit" due to his proclivity for small repairs. Guests would often rely on his aid to open their luggage when they lost their keys, and he was known to spend his time working small projects to keep the Elkhorn in good shape.

More recently, people have reported the presence of a spirit they believe to be that of Uncle John, still hanging around to make sure the property is kept in good repair. One of the Lodge's managers, while walking his rounds one day, found an attic door not only moved but smashed onto the ground. After replacing the door, he again found the same attic door had been moved. A few days later, thinking his attention may have been purposely drawn to the attic, he went up to investigate and found a leaky pipe. After fixing the pipe, the incidents with the attic door ceased. He suspects Uncle John may have moved the door to call his attention to a problem that needed repairs, tending to the maintenance of the Elkhorn in death much as he did in life.

Others have reported ghostly encounters with an old maintenance man who gets upset when someone repairs something without his permission. These stories have likewise been connected to the character of Mr. Fixit Uncle John. Though Rocky Mountain Paranormal maintains a strict policy of neutrality regarding claims of ghostly activity unless and until we're able to personally verify them, we do have to wonder: if these stories *are* true, what might Uncle John think about the substantial renovation project currently underway at the Elkhorn property? Once the Lodge reopens for business in the future, will the stories of the ghostly Mr. Fixit cease, or will the claims increase in frequency?

According to the documentary *Walking History of the Elkhorn Lodge*, Uncle John's spirit still likes to hang out at the dance hall. That also may very well be in line with his character in life. Eleanor's memoirs recount that he was given season tickets to a movie theater, but a third ticket had to be added because he always went with two women. They traveled with him in pairs because he would at times become a bit too affectionate for their tastes.

Charles James is also said to haunt the Elkhorn, though claims involving him are typically more nebulous. On some level it makes a certain degree of sense. If places indeed can be haunted, one would imagine someone like Charlie James would choose to remain at the Elkhorn, where he spent the vast majority of his own short life.

Interestingly enough, the specifically identified spirits associated with the Elkhorn are not limited only to the James family or their close relations or guests. A particularly gruesome story involves Maddie, a young blond girl of approximately eight years of age whose family lived on the property before it was acquired by the Jameses. According to the documentary *Walking History of the Elkhorn Lodge*, Maddie and her brother, while out playing, came across a shotgun. While playing with the weapon and struggling to be the one to hold it, Maddie's hand slipped off the barrel and hit the trigger, causing it to discharge directly into her face. Her spirit is said to still haunt the property, spending most of her time on the back stairs. Toys and dolls have typically been kept on the side of the staircase and occasionally reported to have been moved even when no one was around to move them.

But by far the most prominent of the alleged ghosts at the Elkhorn is Eleanor Hondius herself. One story involving Eleanor began when one of the more recent owners decided to take a nap on the furniture in the lobby. She was awoken by a voice saying "I will not allow that." This is believed to be the spirit of Eleanor objecting to someone napping on her furniture—and indeed, the furniture in the lobby still included the original pieces chosen by the James family.

Figure 12.2. The Elkhorn Lodge lobby. Photo: Bryan Bonner.

A "woman in white" is often seen in the hallways of the main lodge. This spirit is also associated with Eleanor and is sometimes seen in Room 7, now a guest room but historically Eleanor's own room.

There are some additional stories not necessarily associated with any particular character from our historic cast. One guest reported hearing water running in a nearby room in the middle of the night. Knowing he and his family were the only guests in that part of the hotel, he went to investigate and found the room's door was unlocked. The room itself was unoccupied and contained no luggage, but the water was still running. Could a ghost have done this? Or perhaps did a Lodge employee simply forget to shut the water off after cleaning the room or performing a repair? This story was originally relayed to Rocky Mountain Paranormal in an email from someone who claimed his stay at the Elkhorn was in or around 1996. Interestingly, we've heard very similar stories repeated in other sources, including the *Walking History of the Elkhorn Lodge* documentary, in the years since. We can't be certain whether the same witness has repeated the story to other sources, whether multiple such incidents took place, or whether our own report has become the source of a piece of the Elkhorn

Lodge's legend.

Cabin H, one of the guest cabins at the Elkhorn, is the site of several alleged paranormal events. Some of these are fairly commonplace among paranormal lore, such as lights turning themselves on and off. Others are a bit more specific. One guest complained that the beds hadn't been made before his arrival, even though the staff knew that they had been. When they investigated, they found the beds had been completely stripped of their sheets, which had been left in a haphazard pile on the center of the mattress. Ghostly activity? A housekeeping mistake? A mischievous guest trying to scam a free night at the Lodge? It's impossible to say for certain.

Several cowboy spirits have been seen on the property, both inside and outside. One rode up and asked a staff member to stable his horse. The staff member looked away for a moment and when he turned back, the man had vanished. Another staff member also witnessed this event. A guest staying in one of the rooms woke to see a cowboy at the foot of her bed. He also disappeared.

An unnamed homeless man's spirit is said to inhabit the attic of the hotel. According to legend, during his life he kept taking up residence inside the main lodge, and every time he was evicted, he would eventually return and cause a lot of disturbances. Apparently he was prone to drink and fight and generally disrupt the guests and staff. A fight ensued during one eviction attempt during which the man fell down the back stairs and died. His spirit, however, still won't leave and now resides permanently in the attic. He's sometimes called "Attic John," not to be confused with the Uncle John we mentioned before.

Another story involves children of a family staying at the hotel. They remarked to their parents about the nice lady who'd watched over them at night. No such lady was at the hotel at the time...at least not among the living.

The dining room, fit to serve as many as 250 people at a time, is also home to some paranormal claims, including a staff member who saw a group of as many as fifteen people sitting at a table when there weren't supposed to be any people present. When the owner investigated, no one was in there. The staff member who'd originally seen the ghostly party refused to enter the room again.

While not associated with any one spirit in particular, the stables and their hayloft are considered among the most paranormally active locations at the Elkhorn Lodge. According to legend, the tall hayloft was used by law enforcement as a location for hangings. Some of those spirits are supposed to have stayed behind. Additionally, some of the numerous other spirits around the Lodge are said to visit the stables from time to time. Multiple horse wranglers over the years report seeing ghosts sitting in chairs up in the hayloft.

As is the case with any haunted location, there are some of the "standard" claims of paranormal phenomena, including objects moving seemingly of their own accord (one such case involved a wooden gate being ripped from its hinges), lights (or in one instance, gas valves) turning off and on, water running without explanation, and so forth. Because these claims are so ubiquitous in paranormal lore, they're among the kinds of things one is always looking out for, but also difficult to pin down to any particular history or legend.

Finally, the Elkhorn Lodge is home to plenty of stories related to Native American spirits. Like most of the United States, there's a lot of Native—particularly Ute—history within the region. The nearby Oldman Mountain (sometimes spelled Old Man Mountain) has a long history as a sacred location for the Natives. It's one of only three documented vision quest sites in the state of Colorado, where Natives would go on long pilgrimages, fasting for days and awaiting spiritual wisdom. The mountain got its name because one of its rock formations bears a striking resemblance to the silhouette of a Native American chief.

Ghost stories involving Native spirits are commonplace not only at the Elkhorn Lodge but throughout much of the Western United States. However, most of these are not very specific and are often difficult to pin down. People simply report seeing a ghostly apparition of a Native, and generally little other information can ever be obtained, making them notoriously difficult to investigate.

However, one alleged paranormal apparition that is sometimes connected to the Native Americans is the story of the white horse with red eyes. This giant pure white horse with, yes, bright red eyes, has been reported on and around the Elkhorn property by a number of guests and employees. Supposedly it's a kind of protective spirit. Whether it's a ghost horse as some claim or whether it's meant to be some other kind of supernatural entity, it is said to be one of the protectors of this remarkable area.

Additionally, there's a belief among some people that the land on which the Elkhorn Lodge was built is itself a kind of hotbed for paranormal activity. This is associated with the Native American belief in the sacredness of the location (particularly the nearby Oldman Mountain). In modern times, it's been associated with a variety of other paranormal claims that there's a spiritual quality to the land itself for unknown reasons. Some have speculated that there's a magnetic or other physical property in the land that allows it to be a hotbed of paranormal activity.[5]

Some further believe there are graves located on the property. However, these have never been documented. They were originally located by means of dowsing or divining rods, which itself may be enough for many skeptics to dismiss the claim. And the James/Hondius family know nothing about the claim. One of the things we've noticed in our decades of experience is that there are a lot of paranormal claims related

---

5   Similar claims are made, more famously, of such locations as Sedona, Arizona, where the phenomena are associated with so-called vortices (we refuse to use the more popular—but incorrect—plural of "vortexes"), an ill-defined feature that is thought to result in a higher than usual amount of spiritual activity.

to Native American burial grounds.[6] Given that so much of western American history coincides with Native American history, it's unsurprising that a lot of Colorado paranormal claims involve Native burial grounds. An important thing to remember, though, is that over the thousands of years prior to European settlement of what became the United States, pretty much the entire country was, at one time or another, Native American burial ground, so even if these claims are true, it's very difficult to separate the signal from the noise. To our knowledge, no formal research (using more verifiable techniques than dowsing rods) has been done to validate the potential existence of graves on the property. Speaking of formal research....

Our Investigation

Our on-site investigation was a collaboration with The Atlantic Paranormal Society (TAPS) as part of the filming for an episode of the *Ghost Hunters* TV show (Season 2, Bonus Investigation 2). We were contacted by a TAPS member for recommendations of potential Colorado locations for their TV show. As it happened, one of our contacts had suggested to us not long beforehand that we should look into the claims at the Elkhorn Lodge. Putting two and two together, we pitched the idea to TAPS and arranged a joint investigation to be filmed for their show.

Working with the TAPS team was a pleasure and we learned a lot about how paranormal television works. Though we were the ones to introduce the TAPS crew to the Elkhorn Lodge, this investigation was really their show, and we were on hand in order to consult and assist.

Methods in paranormal investigation vary widely from group to group. Some teams use decidedly illegitimate methods, and we're happy to call those out when we see them. Even within the realm of legitimate research, though, there are plenty of different approaches. Our own approach differs significantly from the TAPS approach. In large part this is because TAPS investigations are filmed for television. Not only do the television crews affect the course of the investigation, but whether or not a particular technique makes for good television viewing necessarily affects the design of the investigative protocol. There's nothing wrong with that, but it's a constraint we tend not to face on our own investigations. Therefore, the investigative protocol for this particular research was different from what you've read about in many of the previous chapters. For instance, we typically investigate with the lights on unless the claims of paranormal activity specifically occurred in the dark. TAPS, on the other hand, begin their investigations by turning the lights off, providing a much creepier setting for their television program.

Another factor that affected this investigation was the season. The Elkhorn lodge, at the time of our investigation, still didn't have heat. As such, it remained a

6   This may be partly due to the movie *Poltergeist*.

summer hotel and was closed to the public during the winter months, when we were allowed to investigate. This was both a blessing and a curse. Mostly a blessing. That's because except for the Rocky Mountain Paranormal and TAPS crews and a few Elkhorn Lodge staff, the place was empty. Conducting an investigation in an active hotel, though we've done it and will gladly do it again, is much more difficult. It was also a curse because our investigation in the middle of the winter in an unheated property was one of the coldest experiences we've ever had—and we've had some cold experiences. Temperatures alone were sufficiently frigid to limit our ability to linger in certain areas, particularly the stables.

After the TAPS crew arrived and all proper introductions were made, the current owners and managers provided a tour of the property. Three locations were selected for monitoring during the investigation: the main lodge, the stables (and hayloft), and Eleanor's quarters in Room 7. For ease of reading, we'll use the word "we" to refer both to Rocky Mountain Paranormal and TAPS throughout the discussion of this investigation.

We began our investigation at 7:00 p.m., and concluded at approximately 2:30 a.m. Maybe we could have gone a bit longer, but the bitter cold got to many of the crew members, and it seemed like it was time to pack it up.

Video cameras were placed around the Elkhorn. One camera each was located at the main entrance, the dining hall, the first floor main room, and Eleanor's room. Additionally, a microphone was placed in Eleanor's room.

Throughout the evening, the temperature in the Lodge fluctuated between 20 and 35 degrees Fahrenheit (depending on which room was measured and the time of night). Temperatures outside were in the teens. Our only source of warmth (aside from our own jackets) was a fireplace that had been kept active inside the Lodge. Excursions away from this area had to be brief before we wanted or needed to return to warm back up. The TAPS crew detected some seemingly unusual temperature fluctuations (between 11 and 23 degrees Fahrenheit) using their thermal imaging camera. Though they didn't claim this as absolute proof of paranormal activity, they considered it a plausible explanation.

However, we consider a more plausible explanation to be the presence of numerous windows in that part of the Lodge and poor insulation in the walls. Because the camera wasn't stationary, it was likely measuring the different temperatures on different surfaces. We discovered the insulation issue while investigating odd sounds coming from the exterior walls. Upon looking into these noises we found both the poor insulation and some mice living in the walls.

Sweeps with handheld EMF meters were conducted throughout the investigation, obtaining no anomalous results.

During our investigation of the stables, we noticed that the hayloft above the stables had been converted into a full bar, complete with pool tables and a dance floor.

It was also old and rickety. Of course we loved the creepy ambiance that created, but we question the wisdom of combining such a precarious location with the presence of alcohol. Indeed, even completely sober, we questioned the stability of our footing and integrity of the floors up in the hayloft. While looking around the stables, we heard some strange knocking sounds coming from the seating area in the hayloft/bar. Unfortunately, we were never able to identify the source of the sounds. It was just too cold to spend all night experimenting.

On the way back to the Main Lodge, we saw some glowing eyes in the direction of Oldman Mountain, arguably reminiscent of the stories of the ghostly white horse with red eyes (though the glowing eyes in this case weren't red). We were quickly able to determine that these eyes actually belonged to a passing herd of elk.

One of the creepiest moments occurred in the dining hall. While using the thermal camera, we noticed what looked to all the world like a human figure standing near the main door of the room even though we knew none of our crew should be there. This occurred twice. Eventually, we caught the culprit. Two statues, a lamp, and a table, when viewed through the thermal camera at a particular angle, produced a nearly perfect optical illusion of a human silhouette.

Unfortunately, nothing else happened during our on-site investigation, so we left unable to answer most of the questions. That's a frustrating but common thing in paranormal investigation. Even if one believes there really are ghosts in a location (and we again emphasize our position of strict neutrality on such questions until the evidence is conclusive), one can't force them to manifest at a particular time, so many investigations end much as they began, with collections of stories but very little documentation.

We do have a few additional thoughts concerning some of the paranormal claims surrounding the Elkhorn.

First is the spirit of Maddie, the eight-year-old girl who was said to have died in a firearm accident before the James family established the Elkhorn Lodge. Despite searching, we have been unable to find any records of such an accident. That's not entirely surprising. Something that occurred in the largely-unsettled Estes Valley in the 1870s or earlier may simply not be very well documented.

However, we were able to determine who owned the property before it belonged to the James family and became the Elkhorn Lodge. As you may recall from the History section of this chapter, William James had originally settled a different parcel of land in the region. He traded it for the property on which the Elkhorn now sits with a man known as Reverend William H. McCreery, who remained a friend of the James family long after the transaction was complete.

Though genealogical records (particularly those readily available to the public) are not always complete, we did a bit of digging on Rev. McCreery to determine whether he might have had a daughter named Maddie. As near as we can tell, he did not. The only records of Rev. McCreery's children we could find list his offspring as Mabel, Ida, and Elbert, all of whom lived into adulthood.

With regard to the way the accident itself is described, we have some questions. Trigger guards were a common safety feature on shotguns in the 1870s (just as they are today). Such a rudimentary safety device means it would be difficult, though not impossible, for a hand to slip off the barrel and accidentally hit the trigger while playfully misusing the firearm. Indeed, such accidental discharges are exactly what trigger guards are meant to prevent. Safety switches, however, didn't see widespread use until the 1900s. This combination of factors makes such a tragic accident possible but unlikely.

Does this completely debunk the Maddie story? Not at all. It's entirely possible that someone called Maddie died under such circumstances on or around the property before the James family lived there. However, the origin of the story seems not to have come from any historic records, but from paranormal investigators attempting to decipher what they believed to be spirit communications. We're not going to say such things are impossible or even that these people are necessarily mistaken, but we will say that the evidentiary value of such claims is essentially nil unless and until we can verify that spirit communications themselves are valid, and that's been a decades-long holy grail for Rocky Mountain Paranormal that neither we nor anyone else has obtained.

We have similar reservations about the claims of hangings taking place in the stables and the death of a homeless man now known as "Attic John." Any of these things are possible, of course, but their stories have similarly come from ghost hunters rather than from the historical record. Nowhere in either the public record nor Eleanor Hondius's memoirs is there mention of anything that even hints that these stories are likely to be true. An argument can be made that Eleanor simply wouldn't have written about such horrifying things in her memoirs, which is fair enough, but to claim that the story is therefore true is equivalent to claiming that the lack of evidence is actually evidence for the proposition, and that's simply not enough to convince us.

And that's where things stand. A few of the claims seem unlikely to be true, but the vast majority of them simply remain unknown to us. But there was one other point of interest that came up during our investigation: something mentioned to us in passing that sparked an entirely new branch of our investigation. That will be the subject of the next chapter.

References and Further Reading

Hawes, J. (executive producer) (2006). Elkhorn Lodge (season 2, bonus investigation 2). [TV series episode]. In Hawes, J. (executive producer), *Ghost Hunters,* Pilgrim Media Group.

Hondius, E. E. (1963). *Memoirs of Eleanor E. Hondius of Elkhorn Lodge.* Revised edition (2018). Estes Park, CO: Estes Park Museum Friends & Foundation, Inc. Press.

Lasky, C. (1998). *Ghost Stories of the Estes Valley (Volume 1).* Estes Park, CO: Write On Publications.

Lasky, C. (1999). *More Ghost Stories of the Estes Valley (Volume 2).* Estes Park, CO: Write On

Publications.

Williams, N. K. (2015). *Haunted Hotels of Northern Colorado.* Charleston, SC: Haunted America, The History Press.

Wright, T. K. [executive producer] (2012). *Walking History of the Elkhorn Lodge.* [Film]. Pocket Watch Productions, LLP.

13
The Great Ghost Migration?

There's a strange addendum to the story of the Elkhorn Lodge which also leads into our discussion of Colorado's most famous haunted hotel, Estes Park's Stanley Hotel, in the next chapter. Because the story involves both of these haunted hotels, we gave a lot of consideration to whether it belonged in the previous chapter or the following one, and ultimately decided to add this brief transitional chapter in between.

While wrapping up our joint investigation with TAPS at the Elkhorn Lodge, one of the ranch staff members told us something rather interesting. As this person described it, there was an effort several years prior to write a book containing all the ghost stories of the Stanley Hotel, which was at that time in the process of becoming the famously haunted location it is today. However, they quickly realized that, despite the hotel's haunted reputation and collection of extant ghost stories, they simply didn't have enough to fill an entire book. Therefore, they reached out to the Elkhorn Lodge and "borrowed" some of the ghost stories from their neighboring hotel.

It was an intriguing idea on a few levels. The running joke in Rocky Mountain Paranormal became "how does one move a ghost across town?" And indeed, it raises some interesting questions along those lines. If some of the Stanley's ghost stories were, in fact, borrowed from another venue, then what have all those subsequent visitors and employees who've reported similar tales actually been experiencing? It seems implausible to think that the ghosts themselves, even if completely real, would move house along with their stories. But if they didn't, it would be remarkable for people to continue to interact with those spirits at the new location.

Further, from a folkloric point of view, we're always interested in the development and evolution of paranormal lore, urban legends, and ghost stories. Knowledge of a disruption in the "natural" development of this lore would be quite useful to the more academic side of our paranormal research.

Indeed, understanding folklore is one of the reasons we do what we do. When we sometimes get frustrated by the lack of positive results on many of our investigations, one of the things that keeps us going is the knowledge that we're engaged not only in the search for evidence of paranormal phenomena, but also in the study of folklore. Sometimes people on the more cynical/skeptical side of things question the value of our work. Starting with the assumption that ghosts don't exist, they wonder why anyone should dedicate as much time as we do to their study. Answers to that question are legion, but one of them is that the ghost stories, because they're so popular, actually become a part of the history of a location. In addition to the names and dates of all the people involved at a particular venue, the lore that springs up around it is also important to recognize and remember. It's part of how people understand and interact with their local history, and we ignore it, as we ignore any of our history, at our great intellectual peril.

Therefore, if some of the ghost stories now reported at the Stanely Hotel did indeed originate at the Elkhorn Lodge, it's something we want to know about.

Research into this matter proved more difficult than expected. After years of sporadic research and asking people whether they knew anything about it, we were no closer than when we started. It seemed like a rumor at least some people had heard about, but no one knew whether or not it was true. Regardless, we wanted to know, so we kept working, trying to find examples of ghost stories that might appear, in largely similar form, at both hotels. It was long and largely fruitless work until we took a different approach.

According to the story we'd been told, the ghosts migrated in order for someone to write a book about the history and/or ghost stories of the Stanley Hotel. Therefore, we began looking for all such books we could find, specifically trying to identify the earliest ones, reasoning that if the ghost stories did migrate, it was likely when the first book (or one of the first books, at least) about the Stanley was written. If the book in question had been a later volume, presumably it could have pulled its source material from prior books and the ghost migration wouldn't have been necessary.

We began searching the Library of Congress catalog using search terms including "Stanley Hotel," "Estes Park," and "Estes Valley" among others, to generate a list of early publications on the subject. We then checked the bibliographies of what books we could find to track down any additional works on the topic our initial searches had missed.

Ignoring the most recent publications, we traced the following bibliography of publications on the subject that seemed like they might include the ghost stories in question[1]:

- *Rocky Mountain National Park: A History* by Curt Buchholtz, 1983,
- *Estes Park: A Quick History* by Kenneth Jessen, 1996,
- *Ghost Stories of the Estes Valley: Volume 1* by Celeste Lasky, 1998,
- *More Ghost Stories of the Estes Valley: Volume 2* by Celeste Lasky, 1999,
- *A History and Tour of the Stanley Hotel, Estes Park, Colorado* by Susan S. Davis, 1999,
- *Mr. Stanley of Estes Park* by James H. Pickering, 2000,
- *A Concise History of the Stanley Hotel* by Ron Lasky, 2001, and
- *Stanley Ghost Stories* edited by Susan S. Davis, 2005.

Though numerous books and book chapters on the subject have appeared in the years since these relatively early publications, we considered them irrelevant to our search for the reasons outlined above (though they may still be relevant for collecting the ghost stories in general, they're unlikely to have any bearing on the ghost migration).

Armed with this short list of titles, we set about searching for the ghost stories in question. The first couple of books didn't have what we were looking for. For de facto historians such as ourselves, they're full of fascinating stories, but as we suspected, they don't focus on the Stanley Hotel's ghost stories. Some of the later books either are similarly focused purely on the history (as is the case with Pickering's biography of Mr. Stanley) or tell only the ghost stories already well-established as part of the Stanley's lore (as is the case of Davis's book of haunted tales).

But we think we might have hit the sweet spot with the *Ghost Stories of the Estes Valley* books by Celeste Lasky.

To be sure, they don't match the rumor we were told precisely. Far from being a book exclusively about the Stanley Hotel in which ghosts from the Elkhorn Lodge were migrated, these are actually treatments of the ghost stories of the Estes Park region as a whole. But within their pages, we were able to find stories from both the Stanley Hotel and the Elkhorn Lodge, as well as more than a few whose venue of origin is never specified. And that may be the nugget of truth at the heart of the story, because we also noticed that while most (perhaps all) of the stories in these books taking place at the Stanley Hotel are explicitly mentioned to be from the Stanley, the other venues, including the Elkhorn Lodge, are never mentioned by name. We were only able to identify their source because of our deep familiarity with the paranormal lore of the region.

---

1   For obvious reasons, we ignored works that may be relevant to the history of the Stanley Hotel but which clearly would not contain a substantial treatment of ghost stories. Among those works are volumes about Stanley's steam engines, violins, real estate documents, and so forth.

These books contain some great ghost stories and are well worth reading if you're interested in the paranormal lore of the Estes Park area. Indeed, as far as we have been able to find, these two small books represent the very first publications entirely dedicated to the ghost stories of the area, so their historical importance to paranormal research shouldn't be underestimated. Both for copyright reasons and due to a lack of space, we cannot and will not retell all of the stories in this book. However, in the interest of documenting what we think to be the source both of the rumor that's the subject of this chapter and the history of local paranormal lore, Table 13.1 on the following page identifies the sources, if we know them, of several of the stories in these books. In addition to the stories listed in the table, the two books feature no fewer than fourteen stories whose origin was explicitly listed as the Stanley Hotel and several identified as belonging to private residences or local shops.

Obviously, there are a lot of undocumented stories in these books. They're dominated, perhaps expectedly so, by the Stanley Hotel, but the presence of numerous others including at least two, likely three, and maybe more from the Elkhorn Lodge suggests that these books could be the ones we'd been looking for. What's interesting to us is that the stories from the Stanley Hotel are specifically and explicitly identified as such, but the stories from all the other locations are not. Those omissions may be why the person who originally told us of the ghost migration thought the books were specifically about the Stanley Hotel.

At the very least, we know that the Stanley Hotel at one time offered Lasky's books for sale in their gift shop (that's where at least one of our members bought his copies). We also know that Celeste Lasky's husband Ron Lasky wrote his own (non-paranormal) history of the Stanley in 2001. Their association with the Stanley may explain why that's the only hotel explicitly named in the books. Further, later printings of the books have been revised to include the Stanley Hotel's name in the title.

Now, we're pretty sure (but not 100%) that we've identified the source of the ghost migration story. It's not quite as scandalous as the way we'd been told it originally, but it's still a fascinating bit of paranormal history. We're not going to give up the search entirely—maybe we can find another book that more closely matches the story we were told, and perhaps even some of the stories specifically identified as belonging to the Stanley in the table above could turn out to have been inspired by other locations. We're also going to keep searching to see if we can ever find the sources of some of the unidentified locations in Lasky's books. But for now, we're fairly satisfied with where things stand.

That, of course, means all that's left in this saga is to tell the story of the Stanley Hotel itself. Such will be the subject of our next chapter...and it's a doozy.

Title	Vol.	Listed location	Actual location[2]
The Toy Box	1	Unspecified	Unknown[3]
An Unexpected Guest	1	Unspecified	Unknown
London Bridge	1	Unspecified	Unknown
A Spirited Ringing Out of the Old Year	1	Stanley Hotel	Stanley Hotel[4]
Lost and Found	2	Unspecified	Unknown
The Father's Day Ghost	2	Unspecified	Unknown
A Spirited Horse	2	Unspecified	Elkhorn Lodge[5]
The Overseer	2	Unspecified	Elkhorn Lodge?[6]
Room 237	2	Unspecified	Unknown[7]
The Lady In White – Room 17	2	Unspecified	Elkhorn Lodge[8]
Music Appreciation	2	Unspecified	Unknown
A Housekeeper's Dilemma	2	Unspecified	Unknown
Under It All – Justice	2	Unspecified	Unknown[9]

Table 13.1: Selected stories from Lasky's *Ghost Stories of the Estes Valley* and their sources

---

2   As near as we've been able to determine.

3   There are plenty of stories of toys being moved, including some at the Elkhorn Lodge. However, this is a common theme in paranormal lore, and we've been unable to locate the source of this specific tale, except that Lasky identifies the location as a "lodge." It very well could be a variant on a story from the Elkhorn.

4   There is a similar story of a party of ghosts found in the restaurant at the Elkhorn Lodge, but we have no reason to think this story has been migrated. Stories of ghostly parties in restaurants are not uncommon.

5   This is a variation on the ghostly white horse with red eyes mentioned in the previous chapter. In Lasky's book, the venue is identified only as "the lodge."

6   We've been unable to specifically trace this specific ghost story to the Elkhorn Lodge. However, the ghost described is that of a woman who wears white, sits sometimes in a rocking chair, and grew up at the location, helping to run it once she reached adulthood. That sounds identical to Eleanor's story at the Elkhorn.

7   All we know of this story's origin is that Lasky describes the location as a "ranch" in Estes Park. This could be the Elkhorn, though there aren't enough details to say for certain.

8   Technically, Eleanor's room at the Elkhorn Lodge is Room 7, but the transition from 7 to 17 when this story appeared in print is easy to understand as a typographical error. The remainder of the story is identical to many stories of Eleanor's ghost at the Elkhorn.

9   The story's said to take place at a local hotel that was host to a quilting competition in the 1930s. We've been unable to figure out which hotel that was. However, that was a time during which the Elkhorn Lodge hosted many social functions, so it's not out of the question that it could have been the location in question.

14

Colorado's Most Famous Haunt: The Stanley Hotel

This is the moment many of you have been waiting for. When we give public lectures, some of the most common questions we're asked pertain to what is undoubtedly the most famous haunted location in the state of Colorado: The Stanley Hotel. Located at 333 East Wonderview Avenue in Estes Park, the Stanley is a majestic old hotel whose mythology bridges the gap between fiction and reality. It's been featured, in some form or other, not only in bestselling novels and horror movies but in just about every paranormal TV show there is, and even grand opera.

Rocky Mountain Paranormal has spent a lot of time investigating at the Stanley Hotel. Our earliest investigations date to a time before it was as famous and busy as it is now, and our work there over the years has involved collaborations with film and television professionals, other paranormal groups, authors and historians, magicians, and even several branches of the Federal government of the United States. Buckle up, because these next several pages are going to be a wild ride.

Figure 14.1. The Stanley Hotel. Photo: Bryan Bonner.

The History

Though it officially opened for business on June 22, 1909[1], the Stanley Hotel's history actually begins quite a bit earlier. To understand the developments that led to the construction of this hotel, you need to understand the lives and backgrounds of a fascinating and diverse cast of characters.

Our story begins with a man we met, however briefly, in Chapter 12: The Honorable Windham Thomas Wyndham-Quin, the fourth Earl of Dunraven and Mount-Earl, an Irish nobleman. Born in 1841 and educated at Christ Church, Oxford, Lord Dunraven undertook journalistic and political work following his military service. All of these accomplishments are fascinating and worth reading about in greater detail, but may be of little interest to American readers, so we'll focus instead on the portion of his life beginning when he visited America in 1872 in pursuit of what he'd heard was wonderful hunting in the American West.

During his time on our side of the pond, he befriended "Texas Jack" Omahundro, who would serve his guide on the hunting expeditions. In 1874, having fallen in love with the Estes Park region after discovering it on this journey, he decided he wanted to turn the entirety of the Estes Valley (what is now the city of Estes Park) into a hunting preserve. Sources vary on whether he intended it to be preserved for the public use (as argued by some) or whether he wanted it to be the exclusive domain of himself and his friends and acquaintances from overseas (as argued by probably more sources). The reality may be more complicated. Since Lord Dunraven had no male heirs, the event of his death would cause his property and title to pass out of his immediate family, leaving his three daughters without an estate. A hunting preserve in Estes Park, however, could be just the kind of profitable venture that would leave plenty for his family.

There was just one small problem. Lord Dunraven was not an American citizen, a requirement for the use of the Homestead Act through which he intended to acquire rights to the land. As such, he constructed the English Company, which would later be called the Estes Park Company, and placed his friend, Irish Canadian Theodore Whyte, in charge. Their plan was simple. Hire anyone and everyone they could find to come to Estes Park and file a land claim under the Homestead Act. Each such claim would net a total of 160 acres. All one had to do to secure the title to land under the Homestead Act, in addition to being an American citizen, was to "improve" the land while maintaining residency for (usually) five years. According to Dunraven's

---

1   Not, we emphasize, the Fourth of July, though we've many people repeat the erroneous date. Presumably they like to embellish the mythology by connecting the hotel's birthday to the nation's birthday, or are drawing a connection between the hotel's history and the final shot of Stanley Kubrick's *The Shining*, which we'll discuss in a moment.

arrangement, the people he'd hired would acquire the land and then sell it back to the Estes Park Company, placing it in the hands of Lord Dunraven.

Not everything went perfectly. Just as Lord Dunraven himself had fallen in love with the land, so did some of the men he'd hired to acquire it on his behalf. Some of them refused to sell it back to him, even after taking his payment. When all was said and done, despite a variety of lawsuits from other settlers to the area who considered Dunraven's work to be "land theft," the scheme ended with Lord Dunraven in control of some 10,000 acres of Estes Park, joined, of course, by other (smaller) landowners including the aforementioned James family, the Spragues, the MacGregors, and more.

As William James discovered, the real profit in Estes Park was not in raising cattle, nor in maintaining a hunting preserve. Profit in this area came from tending to the growing flocks of tourists. As F. O. Stanley himself once remarked, "Estes Park is not a place to 'go through.' It has no geysers, no hot springs, no cliff dwellers, and no grizzly bears. But as a place in which to spend the summer in perfect comfort, either in a hotel, or a private cottage, it is unsurpassed."[2] Dunraven came to the same conclusion. On July 9, 1877, he opened his Estes Park Hotel, which the locals came to call the English Hotel. But by 1884, he returned to his home country, never to return to Estes Park in person. Instead, he managed his affairs remotely, acting through his companies. Unfortunately, the English Hotel burned down in 1911, so most of our story won't focus on that. But before we get to the conclusion of Lord Dunraven's story, we need to meet some more characters, whose introduction into Estes Park would completely change the course of local history.

On June 1, 1894, twins Francis Edgar (F. E.) and Freelan[3] Oscar (F. O.) Stanley were born two Solomon Stanley and his wife Apphia French. F.O. and F.E. were two of seven children, all given names of historic, scientific, literary, religious, or family significance. Freelan Oscar and Francis Edgar's names came from a novel by Sir Walter Scott. The other children were Isaac Newton (named for the scientist), John Calvin French (for the French theologian and his maternal family name), Solomon Liberty (for the religious leader; additionally, both Solomon and Liberty were family names), Chansonetta (from the French for "little song"), and Bayard Taylor (for the American poet).

Legend has it that the boys were so alike, even their family dog struggled to tell them apart. The Stanley boys quickly developed a knack for tinkering and an entrepreneurial spirit. By the age of four, they were whittling their own toys. By the age of nine, they were using their father's lathe to manufacture tops to sell to the other children. The same year they began making replacement parts for looms and selling them to the local women in their hometown of Kingfield, Maine.

---

2   Stanley, F. O. (1928). Letter to William E. Sweet. Stanley Museum Archives. Quoted in: J. H. Pickering (2000). *Mr. Stanley of Estes Park*. Kingfield, ME & Estes Park, CO: The Stanley Museum, p. xv.

3   Not Freeland, as is a common misspelling.

As young men, F. O. and F. E. Stanley both became teachers. Both accepted teaching positions at the Western State Normal School in Farmington. But their paths quickly diverged. F.E., upon demonstrating his ability to draw a map of the United States from memory, was accused of cheating and accepted a post at a different institution. F.O., on the other hand, experienced his first illness with tuberculosis around 1871. Though both of the Stanley brothers led the most interesting and productive of lives, it is F.O. Stanley we'll follow more closely in this overview of the history.

F.O. Stanley married Flora Jane Record Tileston, also a teacher, in April of 1876. By all accounts, they remained a loving and devoted couple for the remainder of their lives. And though both loved children, they were not destined to have any of their own. Though we'll not focus on her own biography for the moment, Flora will become a major character a bit later in our story.

Mr. Stanley's businesses were many and varied. While he's probably best known these days for his hotel, it was not the source of his fortune. Neither were his also-famous steam cars. Rather, the Stanley fortune came largely from the photography industry. But these were not his only ventures.

After completing his education. F.O. realized that, following the industrial revolution, drafting tools—triangles, compasses, and the like—were going to be in high demand in educational settings. In 1882, he launched a company producing a complete set of tools which he could manufacture for 25 cents and sell for $1. Orders flooded in and he got his first taste of grander success in business. Unfortunately, he was also soon to get his first taste of disaster in business when his factory burned down. Worse, while the factory was insured, its contents were not. He lost his life savings and was forced to borrow (by some accounts as much as $5,000) from his brother F.E. and attempt to rebuild the business. He was never successful in doing so, though success in other ventures would soon follow.

Meanwhile. F.E. had invented the airbrush—yes, the very same invention that's used by artists to this very day—and started using his invention in the creation of portraits. F.O. would soon join his brother. Together they refined the invention of the Stanley Dry Plate, a photographic process inspired by the airbrush which allowed for the easier and more portable taking of photographs. Many people might argue that the invention of smartphones with high quality cameras allowed for the boom in amateur photography. Others might point a generation earlier to disposable cameras or Polaroids. In reality, photography as we know it would never have existed without the Stanley brothers. Their invention freed photographers from having to prepare wet photographic plates in the field and allowed for the advance manufacture of photographic plates.

The Stanley Dry Plate company quickly saw extraordinary profits, and by 1900, they were receiving more than a million dollars per year in orders. Eventually, several factors converged that would end the Stanley Dry Plate Company. First, F.O. had become ill and was spending his time in Colorado for reasons we'll discuss shortly. Additionally, the photography business was becoming competitive and required a substantial amount of effort to stay ahead of the pack. Probably most importantly, the brothers' interests began to drift. Having achieved their triumph of engineering and business, they wanted to move on to new challenges, and their attention was captured by steam cars. They sold the patents to their dry plate process to one George Eastman of the Eastman Kodak company that dominated film photography until the recent advent of inexpensive digital cameras.

The Stanley Motor Carriage Company was never as profitable as the Dry Plate Company, but it was clearly a labor of love for both brothers. Though they did not produce the first steam-powered automobile in the world, they did produce the first reliable and commercially viable one, and they continued to improve their designs through many different models. The Stanley Steamer, as it's known, holds a lot of records for world's firsts in motoring.

In 1899, it became the first car to climb Mount Washington, a feat which is largely credited with the car's early popularity.

The same year, it became the first car to be ridden by a United States President. Said President was William McKinley who rode the car, driven by F. O., on a tour of Washington, D.C. The President later remarked that the whole time, he was terrified that the car would either explode or run out of control.

In 1903, the Stanley Steamer achieved the world one-mile track speed record.

In 1906, it achieved the world one mile, five mile, and one kilometer speed records. The same year, it would be the first automobile to travel two miles in one minute, and also achieve the world land speed record of 127.659 miles per hour.

Over its 25 years of production, the Stanley Motor Carriage Company produced more than 10,500 cars. As an interesting side note, the Stanleys' airbrush invention ultimately led to the invention of the first fuel injectors (albeit by different inventors for different automobiles), further cementing their importance in motoring history.

Unfortunately, the Stanley Steamer would also claim the life of one of its inventors. F.E. Stanley was killed in an automobile accident while driving his Steamer on July 31, 1918. Reports diverge on the actual cause. The way most people tell the story, F.E. had a habit of driving fast and recklessly, and had numerous accidents, of which this last one was unfortunately fatal. Others claim that the steam engine actually exploded. Though generally safe and well-made, these steam engines did require a great deal of steam pressure, and this was a time before safety pressure release valves had been invented, so explosions were, however rare, certainly a possibility. Either way, the death of his brother was an event from which F. O. would later write that he never fully recovered.

Both Stanley brothers were also passionate violin makers, though this was always more of a hobby than a true business venture on the scale of their other projects. In fact, F. O., probably the superior violin maker of the two (though this is somewhat debatable), made his first instrument at the age of 11 years. F. O. and F. E. just made their instruments for friends and family, though F. O. would later employ his nephew Carlton F. Stanley to produce more than 500 violins which have been played in professional orchestras and are prized by collectors and musicians alike.[4] The Stanley Museum is currently home to the oldest known surviving violin made by F. O. Stanley, which he produced in 1865 at the age of 16 years. In 2016, a sculpture by artists Sutton Betti and Daniel Glanz depicting F.O. Stanley with his violin was installed on the terrace in front of the Stanley Hotel.

In 1903, F. O. Stanley had a severe recurrence of his tuberculosis. Tuberculosis (often historically called consumption, which grim moniker may be more apt than its more scientific name) is a severe, often fatal, communicable respiratory disease we now know to be caused by the bacterium *Mycobacterium tuberculosis*. These days, it's treatable through a long course of antibiotics, though evolved antibiotic resistance is a problem of which we should all be aware. In the early 1900s, though, treatment wasn't so simple. Mr. Stanley's doctors advised him to relocate to Colorado, where the thin dry air might aid his recovery.

Medicine in those days was not what it is today. In fact, this relocation was a common treatment for tuberculosis. It was also a paradoxical one. If someone lived in a humid location near sea level, they'd be told to relocate to a dry place with thinner air like Colorado. On the other hand, Colorado's tuberculosis patients were just as likely to be referred to a location with thicker, moister air as part of their treatment. Remarkably, despite the glaring contradiction in terms, both treatments were often somewhat successful. What was actually happening was that the patient was removing himself or herself from the local community wherein the tuberculosis infection was being passed around like a deadly game of Hot Potato. By relocating, often to a more secluded location but at least to a *different* location, the infection network could be disrupted and the patient might be able to fight off the disease with a diminished risk of reinfection. It turns out that, especially in the infancy of medical science, the best thing one could do was to rest and stop monkeying with a complex biological system one didn't understand.

In any event, the result was the arrival of F. O. and Flora Stanley in Colorado. When they first arrived by train on March 4, 1903, they stayed at none other than Denver's Brown Palace Hotel, which is planned to be the subject of a chapter in a future volume of this series. Stanley's physician in town was Dr. Sherman D. Bonney, who forecast a long stay in the region and Stanley, by then quite famous and well-respected for his inventions and businesses, told the *Colorado Republican* he was enjoying the

---

4   If any reader ever happens to come across a Stanley violin, please ignore that last sentence, pretend the violin has no monetary value, and donate it to the Rocky Mountain Paranormal Research Society.

local climate, expected a long stay, and was beginning to search for "an investment in the way of a home."[5]

Flora Stanley, at the same time, immediately set about making social connections in the new state, attending art exhibitions, meeting the Denver Women's Club, and sitting at a place of honor at a dinner offered by the local chapter of the Daughters of the American Revolution.

In June of that year, the Stanleys finally arrived in Estes Park, on the advice of Dr. Bonney. In characteristic style, F. O. arrived driving his own steam car. Modern readers may not fully appreciate the importance of that decision. Today, it's a simple matter to drive from Denver to Estes Park. Driving by car from the Brown Palace to the Stanley Hotel today takes about 90 minutes and, depending on one's fuel economy, perhaps a little more than two gallons of gasoline. Making the same trip by steam car was a different matter. The Steamer's boiler had to be consistently replenished with water, all of which had to be loaded onto the vehicle in advance. When F. O.'s hired assistant failed to arrive on time, the ill man set about loading his own vehicle with the heavy water tanks and set out on his own, even getting lost once on the way. Still, it must have been quite a sight when the famous stately gentlemen arrived with his marvelous machine steaming and bubbling along the road.

Once again reminding us that the histories of all of these supposedly haunted locations eventually converge if one digs deep enough, not only did the Stanleys stay at the Brown Palace while in Denver, but also at the Elkhorn Lodge when they first arrived in Estes Park.

F. O.'s initial reactions to the region have not been recorded, though one can assume based on what followed that he fell in love with the place as did so many others. Flora, however, wrote an entire article describing how she was "spellbound by the beauty of the scene."[6] The Stanleys decided to stay, and in the months that followed, F. O.'s health began to improve.

To help understand the remarkable character of the Stanleys, it's worth relating two brief anecdotes from this period of time. In October of 1903, F. O. and Flora surprised their niece, Blanche Stanley, on the occasion of her wedding to Edward Merrihew Hallet. No one expected them to brave the winter journey, especially with F. O. still in somewhat poor health. Nevertheless, they arrived the day before the wedding, presenting the bride with gifts of jewelry and money for a new grand piano.

Later that year, Stanley was driving his Steamer through town as he often did, keeping himself under the speed limit of eight miles per hour, when his car was overtaken

---

5   Quoted in: J. H. Pickering (2000). *Mr. Stanley of Estes Park*. Kingfield, ME & Estes Park, CO: The Stanley Museum, p. 52.

6   Stanley, F. (1903). A Tenderfoot's First Summer in the Rockies. Quoted in: J. H. Pickering (2000). *Mr. Stanley of Estes Park*. Kingfield, ME & Estes Park, CO: The Stanley Museum, p. 60.

by an out-of-control horse dragging a wagon to which a small terrified boy clung for his life. Stanley immediately recruited a passer-by into his car and gave chase to rescue the child. They succeeded. One can only imagine how the locals must have reacted.

In the same year, F. O. Stanley designed and built a private home on 8.4 acres of land in Estes Park. Its appearance clearly demonstrates it was designed by the same mind responsible for the Stanley Hotel several years later. The Stanley home is not on the same property as the Hotel, but currently operates as a museum open by appointment only.

Here, the Stanleys' story intersects with that of Lord Dunraven. By the time the Stanleys arrived, the absentee Dunraven was in actual or effective control of the lion's share of Estes Park. Many of the other homesteaders in the area objected to Dunraven's acquisition of so much land. Questions of both legality and ethicality abounded, as did lawsuits. However, in light of the fact that Estes Park turned out, as so many of them discovered, to be more suited to tourism than to cattle ranching, Dunraven's acquisition had its advantages. They might not have liked the way a foreigner manipulated the law to acquire the land, but that so much of it was maintained by a single entity was fortuitous to those who might want to shape the region's development in the future. All they'd have to do was get it out of Dunraven's hands and into their own.

The first to have some partial success in reducing Dunraven's holdings was a local rancher and attorney, Alexander Q. MacGregor, who convinced the courts that many, but not all, of Dunraven's holdings had been fraudulently acquired.

Several people tried to buy the remainder of Dunraven's property but had little success in even getting his attention. He may have prized his Estes Park property, but his attention seemed well taken with other matters on the other side of the world. In 1905, one Guy Robert LaCoste partially succeeded. He visited Dunraven in Europe and returned with a seven-year lease with the option to buy the property for $50,000. Unfortunately, he never found sufficient funding and the lease was dissolved. But in 1907, Mr. Burton D. Sanborn of Greeley, in partnership with none other than F. O. Stanley, purchased the properties for $80,000. Their partnership would eventually dissolve in 1917, but it allowed for Stanley to begin constructing a major resort on the lands that had formerly belonged to Dunraven.

The popular legend often told of the Stanley Hotel is that Lord Dunraven fraudulently obtained the land and then F. O. Stanley hired Mr. MacGregor to steal it back, a tale perhaps bolstered by the fact that the hotel's ballroom is named in honor of Mr. MacGregor. The reality is more complicated and less sensational. While MacGregor did successfully reduce Dunraven's holdings in Estes Park, he did so independently of Stanley's influence. Mr. Stanley's contribution to the project, far from being the "stealing back" of Dunraven's land, was a simple purchase in which Stanley's partner traded some land and Stanley paid in cash. None of the players—Dunraven, MacGregor, or Stanley—ever acted as nefariously as the legend would have us believe, though it probably is true that Dunraven felt somewhat defeated by his rival businessmen when all was said and done.

Mr. Stanley kept himself quite busy in 1907. In addition to breaking ground on his new hotel, he opened an improved road from Lyons, anticipating the need for better roadways to accommodate the predicted influx of tourists and, hopefully, guests for his hotel. Additionally, the Stanley Motor Carriage Company saw fit to produce the Stanley Mountain Wagon, a 12-seat vehicle local resort owners could use to shuttle their guests into and out of the Estes Valley. This led to the foundation of the Estes Park Transportation Company, which quickly obtained a government contract to haul mail into the region.

Figure 14.2. A Stanley Steamer in the lobby of the Stanley Hotel. Photo: Bryan Bonner.

A year earlier, Mr. Stanley had stepped in as president of the Estes Park Protective and Improvement Association (EPPIA). In 1907, he oversaw the construction of a fish hatchery as part of that organization, and formed the committee, along with Enos Mills, that would lead eventually to the dedication of Rocky Mountain National Park in 1915. The very same year, a group of businessmen founded the Estes Park Bank, the first financial institution in the new city, and elected Mr. Stanley, against his wishes, as its president—a role he would fill for 11 years.

He wasn't finished. The next year, he helped to organize the Estes Park Water Company, supplying an essential utility to the community by drawing water from Black Canyon Creek.

Further, because he intended for his new hotel to be the world's first all-electric hotel due to the high cost of coal in 1908, he built Estes Park's first power plant, the Stanley Power Plant, a hydroelectric facility on Fall River. Now called the Fall River Hydroelectric Plant, it still stands, still operates, and has been granted historic protection, though it's no longer the sole supplier of electricity to Estes Park as it was in the early days. Visitors to Estes Park can tour the plant.

In 1909, Mr. Stanley provided land in what was called Stanley Meadows and where Lake Estes can be found today to construct the town's first sewer line. In the years that followed, he would continue to expand the town's septic system.

Originally, the town's first golf course was part of the Stanley Hotel complex. When that proved insufficient, it was Mr. Stanley himself who purchased 120 acres of land near Fish Creek that serves as a golf course even today. However, it's not open for golfers year-round. During the elk rutting season, the beasts and the tourists who come to watch them take over the course. Even the presence of the elk is thanks to Stanley's work. In 1913, he participated in the program to reintroduce the now-famous elk to Estes Park.

And as if all that weren't enough, Mr. Stanley also donated land to form the city's first landfill, stipulating only that it be properly maintained.

Though never the most profitable of his numerous businesses, arguably the crown jewel in Stanley's Estes Park operations was the Stanley Hotel itself. It was not just a place to hang one's hat during travel, but a destination in and of itself. Writing on June 13, 1909 about the hotel's imminent opening, the *Rocky Mountain News* described it as "simply palatial." The hotel was built to Mr. Stanley's own designs, and he spent most of his time at the hotel during its construction, carefully ensuring that everything was done to his specifications.

To say it was an large undertaking would be an understatement. Precise records of the total cost of the project don't exist as Mr. Stanley preferred (wisely, we think) to conduct his business in cash, but estimates range from $150,000 to over $1 million, with Stanley biographer James H. Pickering considering $500,000 to be the most probable figure.[7] Adjusted for inflation, that's roughly equivalent to $16.5 million in present-day figures.

Initially, Mr. Stanley wanted to call the hotel "The Dunraven" in honor of the man who initially acquired the land on which it was to be built. However, the name caused some controversy among the locals who still considered Lord Dunraven's real estate dealings to be at least suspect. When the newspapers—and a petition signed by 180 individuals including all the other resort owners—insisted the hotel should be called "The Stanley," F. O. finally acquiesced.

The Stanley Hotel opened, as mentioned at the beginning of this chapter, on June 22, 1909. Its first guests were a convention of the Colorado Pharmacal Association. Mr. Stanley greeted the attendees personally in Loveland with a fleet of 22 Stanley Steamers ready to take them to the hotel. In a prepared speech to the convention of pharmacists, he welcomed his guests on behalf of "the whole of the Park." These first guests

7   Pickering, J. H. (2000). *Mr. Stanley of Estes Park.* Kingfield, ME & Estes Park, CO: The Stanley Museum, p. 132.

received a discounted rate of $3.50 per day. Standard rates upon the hotel's opening ranged from $5 to $8 per day, substantially more than the standard of the time and in keeping with the Stanley's status as a luxurious mountain resort.

To celebrate the hotel's opening, F. O. presented Flora with a 7 ½ foot Steinway grand piano, which was featured in the musical entertainments of the day, became one of Flora's prized possessions, and still resides at the Hotel. No less a musical talent than John Phillip Sousa tuned and maintained the piano, leaving his autograph on its underside. A later tuner unfortunately mistook the autograph for graffiti and removed it.

Though things typically went well at the Stanley Hotel, this was not universally the case. Any business, and particularly any hotel, has moments of darkness in its history. Tragedy struck in June of 1911 when a chambermaid named Elizabeth Wilson entered Room 217 to light a lamp. Without her knowledge, the gas in the room had been left on and as soon as she struck a match, the room exploded. The blast was so destructive it blew out part of the front wall of the hotel and the floor of the room. Elizabeth fell through the hole to the floor below, breaking both ankles but miraculously surviving the accident. Four busboys also sustained minor injuries. In a testament to Mr. Stanley's good will, as soon as he found out about his employee's injuries he not only covered the entirety of her medical expenses but guaranteed her lifetime employment after her recovery. This kind of care is rare even today and perhaps even rarer at the time, but Mr. Stanley made good on his promise, even securing Wilson's future employment when he would eventually go on to sell the hotel. The repairs to the building cost him approximately $60,000.

Plenty of famous people have wandered the Stanley's halls over the years. Among those are Molly Brown, John Philip Sousa, Theodore Roosevelt, The Emperor and Empress of Japan, Harry Houdini, Jim Carrey, and Stephen King. President Woodrow Wilson was personally invited to stay at the hotel, but he spurned Mr. Stanley's invitation.

By 1926, Mr. Stanley was growing weary of the day-to-day management of the Hotel and his other businesses. Several local businessmen formed The Stanley Corporation, named of course in Mr. Stanley's honor, though not under his control, and purchased the majority of his holdings including the Stanley Hotel. They had grand plans to improve the properties, but their ambitions exceeded their means and they were soon insolvent. Following a complicated series of legal and financial maneuvers beyond the scope of this book, the hotel ended up back in Mr. Stanley's hands in 1929.

Stanley tried his hand at selling the properties again, finding a buyer in 1930 in the person of Roe Emery, who operated the property under the auspices of the Estes Park Hotel Company until 1946. Most of the subsequent owners of the hotel are of little interest to us, except that it's worth tracing the number of times the property has changed hands over the years so we might be able to understand why some things have changed and other things remained the same in the more than 100 years of Stanley Hotel history. The owners[8] have been:

- 1909-1926: F.O. Stanley,
- 1927-1929: The Stanley Corporation,
- 1929-1930: F.O. Stanley,
- 1930-1946: Estes Park Hotel Company (Roe Emery),
- 1946-1965: Abbell Hotel Co (a Chicago-based syndicate),
- 1966-1969: Stanley Hotel, Inc. (a Colorado/Nebraska group),
- 1969-1970: Richard R. Holecheck, Charles F. Hanson, & Carol Hanson Pick (of California),
- 1970-1972: Bankruptcy proceedings,
- 1972-1976: Stanley Properties Trust,
- 1976-1995: Frank and Judith Normali (during a period of complete restoration), and
- 1995-Present: New Stanley Association, doing business as Grand Heritage Hotels.

Before we proceed with the final history of the Stanley Hotel itself, we should sum up our discussion of F.O. and Flora Stanley themselves.

After stepping back from his business interests both in Estes Park and elsewhere, Mr. Stanley remained an active member of the community, summering in Estes Park and wintering on the East Coast. Both F. O. and Flora took a keen interest in works of charity, particularly with regard to children, of which they had none of their own. Stanley biographer James H. Pickering explains that Estes Park residents who'd been children during Stanley's life remembered him as a kind man who'd always help quiz the children on their studies, perform a simple magic trick, or offer wooden gifts he'd made for them in his shop.[9]

In addition to tending to the local children and his various philanthropic (and occasionally political) endeavors, Mr. Stanley spent much of his retirement making his violins. According to legend, many experts struggled to find any difference in quality between a Stanley and a Stradivarius.

While F. O.'s health remained sound for the remainder of his life, his wife was not so fortunate. Flora had suffered a deterioration of her eyesight throughout her entire adult life, eventually reaching a state of near blindness. When examining new

---

8   According to:

Davis, S. S. (1999). *A History and Tour of The Stanley Hotel: Estes Park, Colorado.* Estes Park, CO & Kingfield, ME: The Stanley Museum.

Lasky, R. (2001). *A Concise History of The Stanley Hotel: Estes Park, Colorado.* Estes Park, CO: Write On Publications.

9   Pickering, J. H. (2000). *Mr. Stanley of Estes Park.* Kingfield, ME & Estes Park, CO: the Stanley Museum, pp. 266-7.

paintings together, F. O. would describe the scene to his wife in great detail. The couple were sometimes seen walking together, with F. O. leading Flora so expertly that only those who knew of her condition could tell she was unable to see. Still, by 1935, she seldom left her home.

What she lacked in eyesight, however, Flora Stanley more than made up for in social skill and musical talent. She was known to greet guests at the hotel, even if she couldn't recognize them visually, because she instead recognized their voices. An expert pianist, one of her favorite activities while at the Stanley Hotel was playing her treasured grand piano. She would also sometimes dress as a gypsy and read fortunes to raise funds for the Estes Park Women's Club.

In July of 1939, Flora Stanley suffered a stroke and died 10 days later at the age of 92 on July 25, 1939 from a related cerebral hemorrhage. F. O. was at her side.

The following year, F. O. Stanley himself died at his winter home in Newton on October 2, 1940, ten days after returning from his summer residence in Estes Park. He was 91.

F. O. and Flora Stanley are buried side by side at Sunnyside Cemetery in Kingsfield, Maine.

Throughout most of its history, the Stanley Hotel was a summer resort, which would close in the winters. In fact, the only modern "luxury" the hotel didn't originally offer was heat in the individual rooms. Because it was only used as a summer resort until 1984, it wasn't necessary to have more heat than could be supplied by the fireplaces in the hotel's common areas. However, water heat was eventually added to the rooms in 1976.

Much of the hotel's paranormal legend began in 1974 when one Mr. Stephen King and his wife Tabitha stayed at the Stanley. The Kings had been in Colorado because Stephen was in the process of applying for a job at the *Boulder Daily Camera*. They turned him down and told him he wasn't a good enough writer. And in truth, he probably wouldn't have been nearly as good at the job for which he'd applied as he was at the career he built for himself as a novelist in the following years, though one suspects the newspaper might regret their decision with the benefit of hindsight.

In any case, the Kings arrived at the Stanley Hotel just at the end of the season. All the other guests had left and the employees were preparing to shut down for the winter. Having nowhere else to go and needing a place to sleep for the night, the Kings convinced the staff to keep the hotel open (albeit without most of its services) for one extra night. They stayed in Room 217. That stay famously inspired King's novel *The Shining*, one of the greatest horror novels of the last century.

Such inspiration came from two sources. First, King thought he heard a child calling for his/her nanny in the hallway. As the Kings were the only guests at the time, there should have been no children present. Additionally, he had a nightmare about one of the hotel's firehoses "coming to life" and attacking him. Combined with the unusual experience of being essentially alone in a large hotel, King set about writing *The Shining,* though he clearly indicated in the book that it was not directly based on any of Colorado's real hotels.

King's novel came out in 1977 and Stanley Kubrick released his film adaptation starring Jack Nicholson in 1980. King famously hated Kubrick's movie. For what it's worth, we at Rocky Mountain Paranormal understand where he's coming from. The novel King wrote and the movie Kubrick directed, despite sharing a title and a premise, are *very* different stories, with entirely different themes. However, we think King is wrong. Though not perhaps very good specifically as an adaptation of King's novel, Kubrick's film is a cinematic masterpiece and one of the greatest movies ever made.

But King had different ideas and arranged to film a television mini-series of his own which would be a more faithful adaptation of his novel. That program was actually filmed at the Stanley Hotel, and was released in 1997 on ABC. The mini-series has not been as well received by critics as Kubrick's movie. On some level, that's fair enough—it's hard to compete with the direction of someone like Kubrick. But we at Rocky Mountain Paranormal will be the first to acknowledge that the mini-series, while not the best movie ever, is better than its reputation and is absolutely a more faithful adaptation of King's novel. Plus, since it was filmed largely on-location at the Stanley, it's become a significant part of Stanley Hotel history.

Here, Rocky Mountain Paranormal claims a strange association with both the Stanley Hotel and *The Shining.* One of our associates had booked the Stanley for a wedding. When she visited the hotel to finalize arrangements shortly before the big day, she was told that Stephen King had rented the entire place to film a movie and they couldn't host the wedding. While she was sitting in the lobby trying to figure out what she was going to do on such short notice, someone tapped her on the shoulder and asked what day the wedding was scheduled. That man was none other than Stephen King himself. As it turned out, the wedding was to be on a day when the crew would be filming elsewhere, so our associate was given access to the hotel for the wedding.

Even more interesting, because of the filming, the Stanley had been temporarily rebranded as the Overlook Hotel (the name of the hotel in *The Shining*), making our associate the only person to be married at the Overlook. Because one of our founding members, Bryan Bonner, is also a professional photographer, he photographed the wedding (and may have surreptitiously done a bit of paranormal investigation at the same time—it's not often, after all, one gets to spend some time almost alone in a haunted hotel).

More recently, King's story has also been adapted as an opera by composer Paul Moravec with a libretto by Mark Campbell. It similarly returned directly to the original novel for its source, and debuted in 2016 at the Ordway Music Theater in Saint Paul, Minnesota. The second opera house to feature it was Denver's Ellie Caulkins

Opera House in 2022.[10] Some of our members had the privilege of seeing this performance[11] and all we can say is: if you have a chance to see it, you don't want to miss it.

That wasn't, in fact, the only movie to be filmed at the Stanley. Some scenes from 1994's *Dumb and Dumber* were also shot at the hotel. During the production, actor Jim Carrey was also assigned to Room 217. According to the legend, he ran out of the room in the middle of the night, relocated himself to a different hotel, and refuses to discuss what happened even to this day. We don't know how credible the story is, but it's certainly contributed to the Stanley Hotel's legend.

In recent years, following the popularity of *The Shining* and after several paranormal television shows visited the hotel and spawned flocks of paranormal tourists, the Stanley Hotel has substantially capitalized on its haunted reputation (which we'll discuss in detail in a moment), offering daily haunted tours of the hotel, hosting theatrical seances performed by professional magicians, and even charging premium prices for supposedly haunted rooms (which are often booked up months in advance).

A Brief Tour of the Stanley Hotel Complex

The Stanley Hotel is not a single building. It's a large complex with several buildings, hundreds of rooms, and even a system of underground tunnels. Both to understand the history of the property and because several of these buildings and rooms are invoked in paranormal claims, it's worth taking a moment to briefly tour the facilities.

The main building is of course of critical interest and serves as the hotel proper. If one were to refer to *The* Stanley Hotel, it's the building they'd be talking about. It also features numerous rooms which will become relevant in the stories we're about to tell, so it's worth taking a moment to provide the reader with the lay of the land.

The main hotel's first floor is dominated by a variety of common areas. First of course is the lobby, which originally had white sheetrock walls but these were altered to faux wood as part of the production of *The Shining* miniseries and the change has stuck. Another is the MacGregor Ballroom. It's a grand room suitable for dining for formal events (though daily dinner service is held elsewhere) as well as large parties. Daily bar and restaurant service are provided in the hotel's restaurant called Cascades (formerly the Dunraven Bar and Grille). What used to be the Ladies' Writing Parlor is now the Chrysalis Gift Shop. The Piñon Room, formerly the gentleman's smoking room (despite F. O. Stanley's own abhorrence of tobacco, he understood the needs of his guests), was the location for the bar scenes in King's *The Shining*. There's a Billiards Room, probably much loved by Mr. Stanley himself, who was passionate about

10 It was supposed to be earlier, but the Ellie was closed due to the 2020 pandemic.

11 Rocky Mountain Paranormal member Bob Lewis presented a detailed comparison between the various formats of *The Shining* (including the opera) on his YouTube channel *Phobophile*: https://youtu.be/L26gsD8fVj4

a variety of games. Finally, the Music Room is still home to Flora's beautiful piano but now also doubles as a wine bar.

The second and third floors consists of guest rooms, including the now-infamous Room 217 where both Stephen King and Jim Carrey once stayed. The fourth floor now consists of additional guest rooms. They were once used as staff quarters, in addition to further staff dormitories outside the main building.

The main building along with three others (the Manor House, the Concert Hall, and the Carriage House) comprise the bulk of the Stanley complex.

The Manor House is essentially a miniature model of the main hotel. While the main building had 99 rooms, the Manor House contained 33. However, it was substantially remodeled in the early 1990s with the intent to turn it into a private residence. That never happened, and it's now used for private events and conferences.

The third building, which F.O. Stanley called the Casino, is essentially an entertainment complex and is now called the Concert Hall. In Stanley's day, it was home to many dances and its function has remained largely the same to this day, hosting a variety of musical performances on a regular basis.

The Carriage House was designed specifically to accommodate Mr. Stanley's fleet of Stanley Steamers. It's been used for a variety of purposes in the years since, including functioning as a motel in the 1950s.

On the west side of the hotel complex are several staff dormitories, the former manager's house, the former gate house, the former ice house, a former ice pond, and even a pet cemetery. Despite Stephen King's connection to the hotel, it's unlikely he knew anything about this or that it served in any way as the inspiration for his 1983 novel *Pet Sematary*. In 2013, the pet cemetery was dug up and relocated to a different part of the property, over the objections of local residents, to make room for a wedding pavilion.[12]

Finally, there's tunnel beneath the hotel. The tunnel was originally just a crawlspace but later expanded to the walk-through tunnel it is today. You'll occasionally hear stories of nefarious purposes for the tunnel when people want to "spice up" their ghost stories, but in reality it was simply used as a means for staff to carry objects around, particularly during the winter months.

---

12   Allen, R. (2013). Hotel linked to Stephen King to dig up pet cemetery. *USA Today*. <https://www.usatoday.com/story/news/nation/2013/09/26/hotel-linked-to-stephen-king-to-dig-up-pet-cemetery/2880879/>

Paranormal Claims

Ghost stories abound at the Stanley Hotel. It's at once one of the best and worst documented collections of paranormal lore in the state (and, indeed, probably in the world). Best because there are so many stories that one is never short of material to investigate. Worst because with the abundance of information comes an abundance of misinformation, and it's difficult to separate the original stories from the numerous corrupted retellings that have been written and rewritten, published and republished over the decades.

For example, we were horrified to hear someone recently claim Flora Stanley was a practicing witch. This was not even slightly the case. F. O. Stanley was raised Calvinist and, to our knowledge, he and his wife maintained similar beliefs throughout their lives. The origin of the "witch" claim seems to have been that Flora would occasionally *pretend* to give psychic readings as a form of entertainment, but we suspect the Stanleys would both be aghast to learn anyone took it seriously. Unfortunately, because the ghost stories are so famous, such manipulations of the historic record are commonplace.

Anyway, the end result is that there are far more ghost stories than we can even begin to relate here. Entire books have been written just on ghost stories of the Stanley Hotel. We simply can't repeat all of them. Instead what we'd like to do, first, is simply to point out that there's hardly an area of the Stanley that isn't touched by at least one ghost story, though a few rooms have more than their share. And second, we'd like to provide a taste of some of the more prominent stories and a few of our favorite tales.

By far the most famously haunted room in the hotel is Room 217, so we'll start our discussion there, even though it actually doesn't have the greatest number of paranormal claims associated with it. In addition to Stephen King's report of hearing a child in the hallway while staying in the room and Jim Carrey's experience (whatever that might have been about), there are several alleged hauntings connected to the explosion we discussed earlier. Though the maid Elizabeth Wilson was not killed in the accident, she is no longer among the living and her ghost has been reported by some witnesses in or around the room. Some common events include doors opening or closing, visible manifestations of a spirit, and appearance of a dark circle on the floor where the explosion occurred. Some guests have even reported returning to the room to find their bags had mysteriously been packed or unpacked for them.

Nearby rooms 219 and 222 are frequently the sites of strange and unexplained sounds, as well as doors opening or closing, seemingly of their own accord.

The third floor has a few stories of its own. The most common seem to be doors that don't shut properly. But a few more noteworthy claims have happened in the rooms. In Room 318, some witnesses have claimed to observe a "see-through person." Others have complained that the mattress rotated by itself in the same room. Some guests in Room 340 claimed the bathroom door not only shut by itself, but also locked itself.

But our favorite story (because it's the weirdest) from the third floor happened in Room 317. A bartender was summoned to that room to deliver some ice. When he knocked on the door to make his delivery, a grumpy old man answered the door, annoyed at the interruption, and said he didn't order any ice. The bartender returned to his post and got another call claiming the ice hadn't been delivered. Thinking he simply mistook the floor on which the order originated, he brough the ice to Room 417. The very same grumpy man answered the door, giving the same annoyed response.

But it's the fourth floor where most of the ghostly claims, at least among the guest rooms, originate. Numerous reports come from people who claimed to hear children playing in the halls of the fourth floor or around Room 418. When staff have checked, they've found no children registered on that floor at the time. This is a common paranormal claim in hotels. Every haunted hotel worth its salt has at least some ghostly children in the hallways, often playing with a ball.

One couple stayed in Room 401 and reported seeing the apparition of a man with a bald spot multiple times in their room. Additionally, they smelled the aroma of pipe tobacco. When they later saw a portrait of Lord Dunraven in the hotel, they identified him as the man they'd seen in their room. Indeed, Lord Dunraven is one of the most popularly reported spirits in the Stanley and Room 401 seems to be one of his favorite haunts.

Why Dunraven, who only visited Estes Park briefly and spent the rest of his life in Europe, should haunt the Stanley Hotel is a matter of some speculation. Popular legend holds that he's angry about having been cheated out of his land in the various real estate maneuverings that led to the construction of the hotel. But as we've seen, these dealings were actually much more professional (if not quite amicable) than has been reported in much of the popular press. Still, given he's called *Dunraven*, a ghastly sounding name if there ever was one, had an undeniable connection to the property at one point in time, and is a character in a legend that grew larger than reality, it's unsurprising that he features prominently in the paranormal lore. On the other hand, perhaps his spirit really did just decide to return to the land he'd once acquired. Who are we to question the motives of the spirits?

Moving between the floors, the elevator is also said to be haunted, sometimes operating itself with no one riding it. And ghosts are occasionally seen on the stairs.

Mr. Stanley himself has been seen in a variety of locations around the hotel. On one instance, he was seen staring back at an employee from across the lobby counter. Other times, he's seen wandering the halls or peering out the windows of the old Manager's House.

The MacGregor Ballroom is the site of a few paranormal stories. Some of these involve the explosion in Room 217 because the MacGregor room is directly beneath it and is where Elizabeth Wilson fell after the floor collapsed. But in 1970, a housekeeper preparing the room for a special event witnessed an entirely different kind of haunting: an entire room full of people in period dress, apparently having a party. No one was scheduled to have a party there, but she went about her business. She later said that "something" convinced her to write "Mary Donovan, August 18, 1927" on the mirror in the Concert Hall. She believed that name belonged to one of the spirits she'd seen in the MacGregor Ballroom.

Easily the most famous ghostly resident of the hotel, though, is Flora Stanley, and she's usually found at her treasured piano in the Music Room. Numerous reports exist of people hearing the piano play even when no one is around (often Strauss, Flora's favorite composer) or smelling the scent of roses (Flora's favorite flower) in the room. Some have even seen an apparition, believed to be Flora, sitting at the piano. We can understand this. If we had a beautiful old piano like Flora's Steinway, we wouldn't want to leave it, either.

Please bear in mind that this is nowhere near a complete list of the ghost stories. Not only have we omitted plenty of stories, but even failed to mention some of the rooms that are said to be haunted. One self-professed psychic, upon visiting the hotel, once said she'd never return again, not because the spirits felt threatening, but because there were simply too many of them. We don't usually consider psychics to be the most reliable of sources, but given the number of stories we've collected over the years, we understand where she's coming from.

Strangely, despite the plethora of extant ghost stories, the mythology seems to keep growing. Books of Stanley ghost stories written even just a few years ago don't contain many of the newer stories people are telling these days. Either the property keeps attracting new spirits somehow, or the mythology is simply evolving as people tell and retell the stories. Either way, it's fascinating to watch the claims develop.

Our Investigation

Time to rock and roll. We've spent a lot of time investigating the paranormal claims at the Stanley Hotel, and we have a lot of interesting findings to report. We've been to the Stanley Hotel a number of times, both on formal investigations and, occasionally, just as gusts "keeping our eyes open" while tending to other business.

The first formal investigation began when we were approached by a film production company. They informed us they were going to make a documentary about the ghosts of the Stanley Hotel and asked whether we'd be interested in being the local investigation team for the project. As we'll explain a little bit later, this production turned out not to be quite what we thought it was, but at the time we were perhaps a bit more naïve or at least more eager for television exposure so we readily agreed and worked with the production company and the hotel management to schedule the investigation.

Management at the hotel were excited. They'd long had a haunted reputation and that reputation was becoming more pronounced, so they were keen to capitalize on the publicity a paranormal investigation and documentary could produce. They gave us just about everything we could have asked for. Almost. They didn't quite shut down the entire hotel, so there were some guests there, but not too many, and we were given exclusive access to the areas under investigation. As long as we were careful to keep in mind where the other guests might be, we could investigate without having to worry too much about someone contaminating our data.

We arrived in the afternoon of April 4, 2007 for the investigation. The timing of this is significant. The TAPS crew had been to the Stanley Hotel to film an episode of *Ghost Hunters* not long before. While we participated directly in their Elkhorn Lodge taping, our only involvement with TAPS at the Stanley Hotel was behind the scenes, consulting on a few of the ghost stories and providing introductions to some of the hotel's staff members. But the timing of our investigation was fortuitous in many ways. Because of the television program, interest in the Stanley's ghost stories had been renewed, but the droves of ghost-seekers hadn't yet descended on the hotel. It was a temporal sweet spot. Interest was high enough that the hotel was keen to cooperate, but there were still few enough gawkers around that we were able to conduct our affairs with minimal interruption.

For this first investigation, we were given exclusive access to Room 401 (which you'll recall is one of the paranormal hot spots and rumored to be a favorite location of Lord Dunraven's spirit), Room 418 (which seems to have no more or fewer ghost stories than any other randomly selected room), Room 1302 (part of the Manor House), and the entirety of the Concert Hall.

That's a lot of ground to cover, and we're a small team. We like it that way. Far too often, we've seen groups grow so large that their "investigations" bear greater resemblance to a circus than to a scientific enterprise. Small and dedicated crews seem to be the way to go. However, there's a downside. When an investigation involves four areas spanning three different buildings, it would be nice to have some extra sets of hands. Still, after much discussion, we were able to develop investigative protocols that allowed us to cover all of the area with the following monitoring setup.

Room 401 received two video cameras in the room, one looking at the bed and one toward the closet. Additional cameras were placed to observe the hall and the stairs to the attic. Microphones and thermometers were placed both in the room and in the hallway.

We decided not to station any team members in Room 418. Instead, we placed two video cameras, an EMF meter (within view of a camera) and a digital audio recorder in the room. We then locked the door and left this room alone.

For Room 1302 in the Manor House, we chose the opposite strategy. In lieu of remote monitoring equipment, we stationed two of our investigators in the room to monitor events throughout the evening.

Finally, the Concert Hall received two video cameras in the main hall, one video camera in the hallway, and one video camera at the front door. Microphones were placed in the main hall and at the front door.

Despite its reputation as a particularly "active" room even in comparison to the rest of the haunted hotel, Room 401 was relatively quiet throughout the evening. Members of the production company occasionally popped in to see what was happening, but we'd arranged protocols with them carefully enough that their movements were not enough to hinder the investigation. None of the team members stationed in that room reported anything unusual during the investigation. They took EMF readings throughout the evening, but the results were stable and within the expected range. Temperatures dropped slightly throughout the evening, but not enough to be considered anomalous. This was simply due to the dropping temperature outside the hotel as the evening progressed.

At about 3:30 a.m., a bit of excitement occurred. Investigators monitoring the audio near Room 418 heard a woman. Checking the video cameras, they saw her leave the location and return about three minutes later. Not a ghost, just another hotel guest. The team went back to their normal business. But shortly thereafter, our people heard a very loud noise and started paying attention again. Our hallway video monitoring equipment caught the sight of people exiting the room next door, attempting to steal that room's television. We notified security and went back to our business. We're continually amazed by the weird things that happen during our investigations. Not necessarily paranormal things. Just weird things. Because we put ourselves in strange situations, we have a large and growing collection of odd anecdotes like this.

We were able to solve two of the paranormal mysteries in Room 401. First, we noticed that the closet door would occasionally open or close of its own accord. From time to time, the dresser in the room would also move a bit. Using our seismometers, we were able to determine the source of these movements. The hotel's antique elevator is directly behind the closet. When it passes, it vibrates the floor enough to cause the door to open or close. Additionally, if one doesn't secure the dresser in the room, the vibrations will cause it to "walk" across the entire room in fairly short order.

With regard to Room 418, word reached our ears that a "hypnotist"[13] had been in the room the previous night, seated in a chair at the foot of the bed near the restroom

---

13 Hypnosis is a real phenomenon, but notoriously difficult to understand and its abilities are very often grossly exaggerated. A full treatment of the subject is beyond the scope of this book, but to give just a gist: real hypnosis is characterized by increased focused attention, reduced peripheral attention, and heightened susceptibility to suggestion. It's occasionally useful (though often abused) in mental health clinics. Stage hypnosis, however, in which people are apparently made to do the hypnotist's will, is more akin to an elaborate form of playacting than anything supernatural. On that note, the link between hypnosis and paranormal claims or psychic abilities, while sometimes related in paranormal lore, is tenuous at best.

door. He claimed to sense "something" and said he collapsed when he stood up. The film production company took great interest in this report, interpreting his story as evidence that he'd somehow retained some "energy"[14] because their EMF meters had produced uncharacteristically high readings when they measured him in that location.

We followed up on this idea with our own EMF meters and did indeed detect higher than expected electromagnetic fields in the area—albeit not while the hypnotist in question was present. Some further investigation revealed there is a large power bus that enters the building and passes through the wall behind which the hypnotist was sitting. When the production company's representatives measured the hypnotist, they were not actually detecting his own electromagnetic field but that of the power bus. In other words, he didn't have any stored residual energy in his own body (at least not of any form that could be detected by the EMF meters) but was simply reflecting the field present from the building's electrical wiring.

Aside from this experiment with the EMF detectors, we used only passive video and audio monitoring in Room 418 and obtained no anomalous results.

The Concert Hall provided some interesting investigative challenges. Renovations were in progress at the time of our investigation, and an electrician was working on the building's power panel until later than 2:00 a.m., making it difficult to conduct a full investigation during that time. Instead, we were able to use the time to acclimate ourselves, as much as possible, to the normal sounds of the building. Anyone who's spent some time in an old building knows that these structures often have characteristic pops and creaks. Temperature changes are often the culprit. As the night air cools and the building's heat turns on or off, the building's wood and metal can move slightly, making all kinds of noises. Cars might occasionally drive by. Airplanes might fly overhead. It's useful to become acquainted with these noises so we can differentiate them from anomalous noises later.

Figure 14.3. The Concert Hall (interior) at the Stanley Hotel. Photo: Bryan Bonner.

14   In the contest of paranormal lore and paranormal investigation, energy is an overused and often ill-defined word. The technical definition, from physics, is capacity to do work (which itself is defined as the application of a force to displace a mass). We urge caution when using words like "energy" without fully understanding their technical meaning.

While we were setting up and getting used to the building, the production company related a story that happened on an earlier evening to Grant Wilson from the TAPS crew during the *Ghost Hunters* taping. While in the Concert Hall, a table that had been leaning against the wall "suddenly fell." Additionally, one member of the production team claimed to feel so ill he had to leave the room.

We examined the room for potential causes for both of these claims. With regard to the table, we noticed that the room was being used for storage and many of the tables had been stacked unevenly. While we certainly can't rule out a paranormal explanation (as we weren't present when the incident took place), it seems entirely plausible that a slight vibration (perhaps from a door closing or someone walking across a loose floorboard or something of the sort) could have caused the tables to topple.

The production crew member's feelings of illness are more difficult to account for. Feelings are difficult to explain or measure even with medical professionals on hand to examine the individual in question, and of course that did not take place. However, we noted the extensive electrical renovation project that was underway at the time. Many "hot" wires were exposed throughout the room. It's been established (as we've discussed previously) that high levels of electromagnetic radiation can, under some circumstances, in some individuals (particularly those with certain medical conditions or biological predispositions), cause not only feelings of unease but hallucinations of paranormal or religious experiences. Importantly, we are not saying that's what happened to the crew member. We simply don't know. We raise it only as one possibility based on the conditions we observed in the room.

Things got a bit more interesting for us after the electrician left for the night and we could "go quiet" as we like to do on our investigations. To fully understand what follows, it's helpful to realize a few things about how our investigation was structured. In addition to the video and audio monitoring equipment, two of our team members were stationed in the Concert Hall. They'd made the sound booth above the rear of the auditorium their base of operations.

When we set things up, we were told that hotel staff had only a single key to the front door of the concert hall. They gave us that key and it remained in the possession of our team throughout the evening, so we know no one else could have wandered in during the investigation.

Nevertheless, at about 4:00 a.m., our team members heard what sounded like someone walking around the main hall below where they were monitoring. They stayed silent and listened for a moment as the sound continued to escalate. One of the investigators in the room thought about going to investigate. Writing instead of speaking so as not to contaminate the audio data, this investigator asked the other whether they should end the quiet monitoring phase early and check out the sound.

The written reply was simply, "No."

There's no right or wrong answer regarding which course of action is correct.

The argument for waiting, as they ended up doing, is to simply let things play out for a while and see what happens. Investigation too early can interrupt whatever is going on and result in a missed opportunity to solve a mystery. On the other hand, if one never investigates the anomalies, there's no point in doing what we do. Their ultimate decision was to wait just a little bit longer to see if anything else happened, and *then* go investigate.

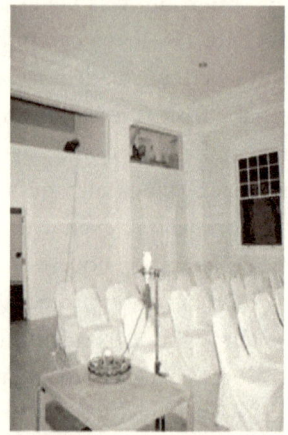

Figure 14.4. The Concert Hall during our investigation. Our crew members monitored from the room above. Photo: Bryan Bonner.

While they were waiting, they heard the sound of a door creaking and slamming shut. Both the initial sounds (which we call footsteps out of convenience even though we don't know for certain that's what they were) and the door closing were caught by our audio recorders. If any of you reading this ever happen to attend one of our public lectures, we'd be glad to play it for you.

At that point, our crew went to investigate. The first thing they checked was to see if anyone else was in the Concert Hall. No one was. They also checked every single door in the building to determine which of them might make the creaking sound they heard. Only the main entrance door made that sound. But the only key to that door was in our possession. Furthermore, one of our cameras was pointing directly at that door, and it did not move when the sound was heard (nor any other time except when our people were using it to enter or exit the building). No further anomalous sounds were heard throughout the evening.

To this day, we don't know what caused those sounds.

We both love that and find it frustrating. We are obsessed with mysteries, and to have a mystery like this is a continuing source of excitement for us. But we approach mysteries with the goal of solving them, and we're usually pretty good at it, so it's frustrating when we have to leave some questions unanswered. It's that quest that keeps us going.

Nothing else of interest occurred within the Concert Hall during our investigation. EMF readings were taken throughout the evening and, allowing for the bad wiring and renovation project, none of the measurements were anomalous.

The Manor House also gave us some good results that evening, albeit results of a different character. A few of our team members were headed up to Room 1302 in that building. The upper floor of the building was pitch dark. Apparently we'd forgotten to turn on the lights. No problem. Though we typically don't investigate in the dark, neither are we particularly afraid of it. Our group marched on in darkness and entered the main lobby of the second floor.

Immediately, the first person to step into the room screamed and stepped back.

This is uncharacteristic. While we can appreciate the fun of running around in the dark and scaring yourself or your friends, that's not what we're doing on an investigation. When we're working, we're engaged in a sober and rigorous search for answers. Plus, our people have seen just about everything on these investigations and we don't rattle easily. Screaming is pretty close to never part of the process.

Of course we immediately asked what happened and the team member in question explained that they'd heard something move up ahead. Worse, they saw glowing eyes staring at them from the same direction. Again, not the sort of thing we normally encounter, and we all got pretty excited.

But in this case, we caught the culprit.

Figure 14.5. We caught the culprit in the Manor House. Photo: Bryan Bonner.

As soon as we turned on the lights, it was immediately clear that the source of the disturbance was a raccoon who'd crawled in through the slightly open window. Many of our crew are animal lovers, so this pleased us *almost* as much as if there'd been a ghost in the building. He was a friendly and curious little fellow who probably just wanted to see if there was any food inside the building. We named him Casper. After looking around for a moment, Casper wandered back outside and joined several other raccoons.

We wrapped up the investigation fairly satisfied with the work we'd done. We managed to solve a few mysteries, experienced some others we couldn't explain, and most importantly, got to spend the night in a remarkable historic hotel. Not bad for a day's work.

Little did we realize at the time, our work at the Stanley Hotel was just beginning.

As it turns out, the production company involved in the investigation was actually working on something rather different from what we initially thought it was. The documentary they eventually released bore strikingly little resemblance to our investigation. It came out in 2008 and is called *The Stanley Effect: A Piezoelectric Nightmare!* (the exclamation mark is theirs, not ours). It's worth reading the production company's own description of the "documentary," which we present in its entirety below:

> Guests of the Stanley Hotel in Estes Park, Colorado sometimes report odd impressions of unseen children and disembodied voices that can be recorded. Some visitors claim that these and other perceptions can become a physical reality that defies explanation. Watch as we present evidence of the unusual properties of the grounds [sic] ability to hold and release energy in ways that are unimaginable. Some of Colorado's finest researchers and various U.S. Government agencies present cases for and against the claims of the earth's ability to record and play back sights, sounds and perhaps even emotions. The Piezoelectric effect of the crystals seem to be amplified by the very power plant that F.O. Stanley built to electrify his grand hotel. Did Stanley unintentionally create the perfect environment for capturing and releasing the essence of events that occur in his magnificent hotel?[15]

There's a lot to unpack here. First of all, it needs to be stated that some of our members appear in this documentary, because we were filmed as part of the production before we knew exactly what it was really going to be about. To their credit, the filmmakers included a substantial portion of our interviews, including several clips that argued against a variety of specific paranormal claims. They gave us our due. However, the film on the whole remains strongly biased in favor of the paranormal ideas, and one idea in particular.

Some of the ideas in the documentary are ones we'd heard in passing before, but the release of an entire film devoted to them got us interested in digging in a bit further. The fundamental claim is that the Stanley Hotel is built on ground rich with quartz, magnetite, and limestone. These minerals, through the piezoelectric effect,

---

15   Murphy, B. (producer) (2008). *The Stanley Effect: A Piezoelectric Nightmare!* [documentary]. Ghost Town TV.

can produce an environment hospitable for ghostly manifestations (specifically, they claim that these minerals somehow "record" and "play back" events and emotions that occurred in the past, making this what's referred to in paranormal lore as a passive haunting. According to the lore, a passive haunting is like rewatching a video recorded event from the past. An active haunting is a ghostly being with conscious agency. We're aware of the distinction, but we tend not to worry too much about it in our investigations. We figure we'll first try to document *if* there's actually a haunting, and only then will we worry about categorizing it.

Let's begin with the piezoelectric effect. In physics and materials science, this is defined as a property of certain minerals by which materials can produce an electromagnetic field in response to applied mechanical stress. Basically, when you squeeze a piezoelectric material, it generates electricity. This is a real, well-documented, and well-studied effect in physics. Many modern electronic gadgets and gizmos make use of it. A good and familiar example is a "click style" lighter. When you press the button, you accomplish two things. First, you release a small amount of butane gas. Second, the button compresses a small piezoelectric crystal, producing an electric spark that ignites the gas.

That brings us to the second issue. It's claimed that the haunting at the Stanley may be due to a piezoelectric effect resulting from the presence of quartz, magnetite, and limestone. A good question to ask would be: are these materials actually piezoelectric?

Quartz is the easiest one. It absolutely does exhibit the piezoelectric effect. In fact, quartz crystals were among the first materials demonstrated to possess piezoelectric properties and remain the best known (and probably most commonly used) of all such minerals (at least among the naturally occurring ones; some ceramics and synthetic materials also exhibit piezoelectricity).

Limestone is also a fairly easy case. Sedimentary rocks like limestone are not known to exhibit piezoelectric properties. The same is generally true of magnetite, a ferromagnetic iron oxide. It is both attracted to magnets and can become magnetized itself, but while magnets (obviously) possess a magnetic field, they don't possess an electric field. Piezoelectric effects can be observed in some materials of which magnetite is a component in the laboratory, but it does not appear to exhibit piezoelectricity in nature.

So, some of the minerals under consideration seem irrelevant. But if there is an abundance of quartz crystal under the Stanley, perhaps we don't need to write off the piezoelectric effect entirely.

This all got us thinking, and we wanted to know where this idea came from in the first place. The earliest report we could find claiming that the haunting at the Stanley could be related to its geology came from the *Ghost Hunters* television program. Because we're acquainted with the TAPS crew, we called them up and asked where they got the story. Turns out, their research wasn't quite as deep or rigorous as ours. They'd gone to the Stanley Hotel's gift shop, and whoever happened to be working at the sales counter that day told them about the alleged geological effect. Not exactly the most reliable of sources.

However, it got us interested in the problem. While piezoelectricity is a known phenomenon, we can't think of a good way to determine whether it's in any way connected with ghosts. That claim seems rather beyond our (or anyone's) ability to either verify or debunk. What we could do, though, is determine whether or not those minerals actually exist on the Stanley grounds. We may not be able to answer the paranormal question, but we can at least scientifically look at the natural phenomena at the paranormal lore's periphery.

So we started writing letters to determine whether a geological survey had been conducted on the Stanley's grounds. We first assumed that the United States Geological Survey was responsible for soil surveys, but it turns out instead to be under the jurisdiction of the United States Department of Agriculture.

We sent them a letter and just laid it all out. We explained who we were and that someone had made a paranormal claim that invoked geology, and asked whether they could provide any data regarding the geology of the Stanley Hotel grounds.

A representative of the U.S.D.A. responded by writing, "Apparently, ghost hunters are interested in our data now. It is the most unique and interesting request I have received in my 30-year career!"[16]

Unfortunately, we were informed that no soil survey had been done anywhere near the Stanley grounds. And anyone who knows anything about geology will tell you that a survey from even a relatively short distance away will have very little bearing on the geology of the site in question. Short distances can have a big effect on local geology. However, our new contact was able to inform us that a geomagnetic satellite survey had been done several years earlier.

"I reviewed the Aeromagnetic data for Colorado," he wrote. "Aeromagnetic surveys detect changes in the earth's magnetic field. The magnetic field is strengthened by the presence of metallic components in bedrock and more so by the presence of minerals with magnetic properties such as magnetite. There is nothing unusual about the aeromagnetic data in the area of Estes Park as compared to the general area of the Rockies. I hope this helps. So at this point it looks like the magnetite (or anything magnetic) in nature is slowly getting ruled out, but I will continue working."

That counted for something. At the very least, it made high concentrations of magnetite seem unlikely. But we weren't entirely satisfied. A satellite survey is not the same as a soil survey, and could do little to detect piezoelectric (as opposed to magnetic) effects. Despite our diverse scientific backgrounds, though, Rocky Mountain Paranormal simply didn't have the resources to conduct a soil survey of our own, so we were prepared to accept what little information was available and leave some question marks in our case files.

---

16 We aim to please.

Things changed a few months later. As it happened, our correspondence had been making its way through the scientific departments of the Federal government without our knowledge, and a response eventually made its way back to us. Reading government correspondence is fascinating. The entire history of forwarded messages and replies had been maintained within the messages, so we were able to track our inquiry bouncing around the government halls. By the time it came back to us, the subject line on the message had been changed to "Ghosts in Our Soils." We're particularly proud to have gotten government scientists to send official correspondence under such a heading.

The gist of the letter that came back to us was this: because a soil survey had never been done at the location and the government had a scientific interest in performing such a survey[17], they wanted to know whether we could get their scientists into the Stanley Hotel for a study.

Allow us to repeat that for emphasis. The Federal government of the United States was asking us—a little paranormal research society from Colorado—whether we could get them access to a historic hotel for a soil survey.

Well, as we all learned from *Ghostbusters*, "when someone asks you if you're a god, you say yes."

We reached out to the owners of the Stanley Hotel and worked on coordinating everyone's calendars to schedule a soil survey. Because we were involved in the planning, we got to participate in the research when a team consisting of ourselves and scientists from the United States Department of Agriculture (USDA) and its subagencies the Natural Resources Conservation Service (NRCS) and the National Soil Survey Center (NSSC). The survey took place over the course of two days in 2008 and involved a team of several scientists and a lot of specialized equipment, some of which (and some of whom) had to be flown in from out of state. We spent *so* many of your tax dollars over those two days, it's not even funny. You're welcome.

The survey itself consisted of ground penetrating radar to "look" at the various geological formations underground, electromagnetic soil surveys which used electromagnetic induction to determine the soil salinity (saltiness) and other electromagnetic properties, as well as chemical soil analysis to determine the soil content. Over the course of the two days, the scientists took readings on the main property of the Stanley Hotel as well as the area near the old ice pond and (now relocated) pet cemetery and a nearby ranch.

---

17 They weren't just excited to play ghost hunters for the day (though the scientists did seem to enjoy that part of the project as well). The Natural Resources Conservation Service was completing a soil survey of the surrounding areas at the time and were interested specifically in finding the patterns of depth to bedrock in the area.

Figure 14.6. Surveying equipment on the grounds of the Stanley Hotel. Photo: Bryan Bonner.

A couple interesting things took place during the survey. First, when the soil scientists toured the tunnel under the Stanley Hotel, staff members were keen to tell them some of the history and ghost stories of the property. The scientists, though, were much more interested in the tunnel itself, because it had been cut directly out of the ground and represented a very rare opportunity to see that kind of untouched soil.

Secondly, because we're never too serious about our work to engage in some shenanigans, we planted a mock "crystal skull" from our collections in the soil and photographed one of the soil scientists "discovering" it.

Figure 14.7. A "crystal skull" we "discovered" on the grounds of the Stanley Hotel. Photo: Bryan Bonner.

Crystal skulls are quartz crystals carved into the shape of human skulls. They were originally claimed to be pre-Mesopotamian artifacts, though scientific scrutiny has revealed all of them we know about to have been made in the 19th century or later. They're associated with a wide variety of paranormal claims, conspiracy theories, and a substandard *Indiana Jones* movie.

We thought it would be funny if we could start a rumor that one of the lost crystal skulls had been found at the Stanley Hotel. Alas, none of the publications that report on paranormal phenomena and conspiracy theories ever picked up on our photograph. It's probably for the best, though we still think it would have been hilarious.

The official report of the survey was distributed on October 15, 2008 and marks if not the first time that paranormal investigators have been included as research participants in a government scientific report then at the very least one of the few times such a thing has happened. Which brings us to the moment of truth. What were the survey's findings? The entirety of the report consists of a lot of technical measurements and is far beyond the scope of this book, but to summarize as succinctly as possible: just dirt.

To provide just a little bit more detail, they found that the soil consists primarily of a substance called schist, a rock formed by dynamic high temperature and pressure metamorphism. The strain aligns the mica, hornblende, and other minerals into thin layers, with such layers consisting of at least 50% of the mineral grains. The distance to bedrock in the area ranges from surface level to approximately 27 inches. No large deposits of quartz or magnetite were found, nor was any evidence of anomalous piezoelectric or magnetic activity.

So where does that leave us? The way we see it, we have no idea whether or not the Stanley Hotel is haunted. We also have no idea whether piezoelectric effects have anything to do with paranormal activity. But we can confirm at least that in this particular case, the story of piezoelectric effects causing the paranormal activity at the Stanley Hotel is false. Maybe it's haunted for different reasons and maybe other places with large quartz deposits have hauntings due to that property—we'd have to investigate to find out. But this particular story simply isn't true.

We prepared a report of our findings, appending the results of the soil survey to the description of our own paranormal investigation some months earlier and distributed it to everyone we could think of with a stake in the matter. We don't know the hotel staff's initial reaction to the report. However, to their credit, the guys at TAPS were quite interested in what we had to say and stopped repeating the now-debunked portion of the claim.

In the scheme of things, it made little difference.

In 2010, Zak Bagans and his crew for the *Ghost Adventures* television program filmed their visit to the hotel. In their episode, both Bagans and the hotel staff member giving him a tour of the property repeated the claim that the hotel has ghosts because of the large quartz deposits.[18]

As recently as 2019, Crystal Ro, writing for that great bastion of intellectualism and journalistic integrity, *Buzzfeed*, wrote an article repeating the same debunked claims.[19]

We're not asking people to stop telling the ghost stories. Even we like to repeat the stories, as you've already seen. But it doesn't seem like it should be too much to ask that people stop repeating portions of the stories that are demonstrably false.

Still, we're used to the fact that folklore often trumps science when it comes to getting headlines, making television shows, or telling spooky stories. We're disappointed but not surprised.

Our involvement at the Stanley didn't end with the soil survey, though.

Some years later, we were given the opportunity to conduct a follow-up investigation when we were contacted by an author who wanted to write about the hotel. We'll not name the author because the book deal ultimately went south, but the investigation that resulted had a few interesting moments.

We arrived in the afternoon and set up, planning to start the investigation at about 7:00 p.m. This time, we set up a base of operations at one end of the Billiards Room and established video, audio, EMF, temperature, and vibration monitoring in the Billiards Room, the MacGregor Ballroom, the Music Room, and the front patio. We also had access to Room 401 (again) and the Carriage House for additional investigation.

It turned out to be a very different experience this time around. The paranormal tourists who now account for a substantial portion of the Stanley's business had arrived by then, so the hotel was much busier than it had been in the past. When word got around to the guests that the paranormal investigators were in town, many of them wanted to see what we were up to, and the constant interruptions made conducting a proper investigation all but impossible, so much of the evening ended up being essentially a "Q&A" session for the hotel guests rather than a paranormal investigation (not that there's anything wrong with that; we like answering questions). Many of the guests wanted to share their "orb" pictures (which we'll discuss in detail in a later chapter) or relate their personal experiences. On a normal day, hearing people's ghost stories is one of our favorite things. When we're trying to conduct an investigation, it does make things a bit confused.

We abandoned any hope of being able to investigate in the matter to which we're accustomed, but we're nothing if not adaptable. For part of the evening, we gave several tours of the hotel to people who were curious about the ghost stories. Throughout the evening, we had issues with people sneaking into the locations we were trying to

---

18   Bagans, Z. (executive producer) (2010). Stanley Hotel (season 4, episode 5) [TV series episode]. In Z. Bagans (executive producer), *Ghost Adventures*. MY Entertainment.

19   Ro, C. (2019). 17 Interesting Facts About The Haunted Hotel Stephen King Stayed In When He Came Up With "The Shining." *Buzzfeed*. <https://www.buzzfeed.com/crystalro/stanley-hotel-shining-doctor-sleep>

investigate because they wanted to see the "ghost hunters."

When we would take our breaks, we'd sometimes retreat to Room 401. People regularly knocked on the door and began talking to the ghosts. Others hovered outside telling each other "this is Jason's room," in reference to one of the TAPS members who'd visited the room on television.

On a related note, during an episode of our horror themed *Do You Like Scary Movies* podcast[20], our friend, the filmmaker Zack Beins shared a great idea with us. If you're planning to stay at the Stanley Hotel, book yourself one of the famously haunted rooms—401, 217, etc. You may have to book as much as a year in advance for some of those rooms, but it's worth it. When you get there, hang out in your room, sitting silently by the door in the middle of the night. It won't be long before some ghost enthusiasts walk past and whisper at each other about the ghost stories while standing right outside the door, possibly posing for photographs. This is your opportunity to pound on the door as loudly as you can and emit the most ghostly sounds you can think of. You'll have a good laugh and those people will have the "paranormal experience" of a lifetime. We haven't had the opportunity to try it yet, but it's on all of our bucket lists.

A common story the guests told us was that there was a "vortex" (an ill-defined term, though typically related in ghost lore to orb photographs) near the stairs on the second floor. One group claimed to have a video of a light that was changing color near a door at the vortex. We were able to determine that there's a power main in the area which causes a substantial electric field at the base of the stairs. People who obtain high EMF readings there have ascribed their results to the "vortex" even though it's actually just the hotel's electrical system. The flashing light, meanwhile, was caused by a reflection from a door handle. It seemed to change color due to the low-quality sensor in the inexpensive camera the individuals were using.

While we were looking around that location, other people told us they'd smelled cigar smoke in the same area. Of course we investigated, and were able to determine they were indeed smelling smoke, but not from a ghostly cigar. A guest in a nearby room was attempting to cook bacon on a hot plate and burned it. Ruined bacon, to us, is far more terrifying than any ghost. Don't burn such a magical food!

This particular anecdote also contains an important psychological lesson. Smoke from a cigar smells almost entirely unlike smoke from overcooked bacon. If one is predisposed to think they may smell cigar smoke due to ghostly activity (which is part of the hotel's paranormal lore), though, it's easy to mistake even a familiar smell (bacon) for whatever we're psychologically primed to think we're going to experience.

Around 2:00 a.m., we found a substantial leak coming from the ceiling in the Billiards Room. Someone in a room upstairs had left the bath water running and left the room. We reported this to the hotel immediately.

---

20 It's at www.doyoulikescarymovies.com as well as all your favorite podcast services. We'd appreciate if y'all would give it a listen, and we think you'll enjoy it.

The most significant story of the investigation happened earlier in the evening after some of the guests asked us about an incident they'd seen on the *Ghost Hunters* program. In one of their episodes on the Stanley Hotel, there's a scene in which TAPS member Grant is seen at a large wooden table. In the scene, the table is seen to move, which is attributed in the show to ghostly activity.[21] Since the table in question was in the Carriage House and we had access to that portion of the property, we headed over there to do a bit of investigating.

While we were there, incidentally, Casper the raccoon made another appearance. This time the family of raccoons stayed outside the window, though.

We determined that the table in question, though large and heavy *looking* is actually a very light pressboard table which is easy to move. Indeed, we could lift it entirely off the ground with just a couple fingers. Careful analysis of the *Ghost Hunters* video reveals that Grant accidentally bumped the table with his right leg, causing it to move and inadvertently scaring himself. Because we had a few minutes to spare, we filmed a comedic recreation of the scene and published our recreation and explanation online.

Fortunately for the Stanley Hotel but unfortunately for those of us who'd like some time to research in silence, business at the hotel seems to be booming these days, so we haven't been able to return for further investigation in the last several years.

References and Further Reading

Davis, S. S. (1999). *A History and Tour of The Stanley Hotel: Estes Park, Colorado.* Kingfield, ME & Estes Park, CO: The Stanley Museum.

Davis, S. S. (2005). *Stanley Ghost Stories.* Kingfield, ME & Estes Park, CO: The Stanley Museum.

Garris, M. (director), & King, S. (writer) (1997). *The Shining.* [TV series]. Lakeside Productions & Warner Bros. Television.

Hawes, J. (executive producer) (2006). Stanley Hotel (season 2, episode 22). [TV series episode]. In Hawes, J. (executive producer), *Ghost Hunters,* Pilgrim Media Group.

King, S. (1977). *The Shining.* New York: Doubleday.

Kubrick, S. (director) (1980). *The Shining.* [film]. Warner Bros.

Lamb, K. (2016). *Ghosthunting Colorado.* Covington, KY: Clerisy Press.

Lasky, C. (1998). *Ghost Stories of the Estes Valley (Volume 1).* Estes Park, CO: Write On Publications.

Lasky, C. (1999). *More Ghost Stories of the Estes Valley (Volume 2).* Estes Park, CO: Write On Publications.

Lasky, R. (2001). *A Concise History of The Stanley Hotel: Estes Park, Colorado.* Estes Park, CO: Write On Publications.

---

21   Hawes, J. (executive producer) (2006). Stanley Hotel (season 2, episode 22). [TV series episode]. In Hawes, J. (executive producer), *Ghost Hunters,* Pilgrim Media Group.

Murphy, B. (producer) (2008). *The Stanley Effect: A Piezoelectric Nightmare!* [documentary]. Ghost Town TV.

Pickering, J. H. (2000). *Mr. Stanley of Estes Park.* Kingfield, ME & Estes Park, CO: The Stanley Museum.

Williams, N. K. (2015). *Haunted Hotels of Northern Colorado.* Charleston, SC: Haunted America, The History Press.

PART TWO
Private Residences

Though the public venues we've discussed so far (and will continue to discuss in future volumes of this series) are arguably of the greatest general interest, they don't represent the entirety (perhaps not even the bulk, though it varies from year to year) of our work. On a regular basis, people write to us or approach us at one of our public lectures and ask for our assistance with their personal paranormal experiences.

Private cases such as these come in all shapes and sizes. Some of them never proceed past some initial email discussion or perhaps a quick forensic analysis of a single photograph, while others may involve substantial on-site research on par with the kind of work we've done in some of the public cases.

For this second part of the book, we'd like to take you through a tour of some of these private cases, but there are a couple of caveats to get out of the way before we do so.

First, though we're attempting to provide a complete record of our case files in these books (though it will take several volumes to do so, especially since we're still actively conducting research on new cases), we're not going to document every communication we ever receive. Many cases simply don't go anywhere, and while we maintain records of these cases in our private files, we wouldn't presume to bore the reader with repetitive details of cases we were never able to fully investigate. On the other hand, neither are we presenting only the most outstanding and extraordinary cases. We do want to provide a reasonable overview of our activities.

Second and most importantly, because these are not public venues or public figures, we will not identify the individuals into whose homes and lives we've been invited. Most of the time, we can accomplish this simply by omitting their names or assigning pseudonyms. Every once in a while, some feature of the case narrative itself might be potentially identifying. When that happens, we will disguise the individuals' identities in a variety of ways, but never in such a way as to alter the relevant data we're trying to present.

With regard to photographic evidence, we have similar concerns. Occasionally, we may have a photograph that's restricted enough in its contents (or can be cropped to be so restricted) that it can illustrate the relevant scene without risk of identifying or embarrassing the client(s). When that's the case, we'll use genuine photographs. The rest of the time, we'll use artist's renditions to illustrate the cases.

On occasion, we've mentioned our work to people of our acquaintance who aren't involved in the same kind of research. Reactions to the claims made by our clients are not always positive. It's easy to dismiss some of these claims as being those of crazy people. But that's not been our experience. While there certainly are some people who suffer from a variety of mental health conditions ("crazy" is not a technical term) and occasionally they cross our path, most of the people who reach out to us are fundamentally sane and normal. Paranormal lore is a part of our culture and plenty of reasonable people believe in it.

Often, perfectly sane and normal people who simply happen to have a belief or interest in paranormal phenomena have some unexplained event in their lives, and we consider it our duty to help them understand what they're dealing with. If we can find a natural explanation so they don't have to be afraid of their own homes anymore, we consider it a good day's work. On the other hand, if we could document a legitimate paranormal event to validate what they thought they were experiencing, we'd also consider that a good day's work.

In the pages to follow, you'll meet a diverse cast of characters, and we hope you enjoy their paranormal stories and our research into the same.

Note: though the chapters in Part One were divided into sections for history, paranormal claims, our investigation, and further reading, the private cases will not be. Most of these residences don't have much history to speak of, and further reading on these particular cases is typically nonexistent. Instead, we'll present the entire story in narrative form.

ESSAY
The Life Cycle of a Private Paranormal Investigation

In the introductory material of this book, we walked through a rough sketch of what our investigations are like. As you've read through Part One, you've seen it in action at a variety of public venues. You may have also noticed that, despite some over-arching themes and basic protocols, all of our investigations are a little bit different. Because each claim is unique, each investigative protocol is also unique.

Public and private investigations share most elements in common, but there are some important distinctions. First, we've noticed some difference in terms of the claims that are made. The vast majority of our investigations of public venues have involved claims of ghosts. Private investigations have plenty of ghostly claims as well, but we've found it much more likely for private individuals to fear they're being tormented by a demon.

Reasons for this are numerous and probably no one of them is the sole right answer. One thing we suspect may be going on is that if you own a public venue, a ghost story may attract curiosity seekers while a claim of demonic possession might scare most paying customers away. On the private side, a lot of paranormal events coincide with major upheavals in life, and the negative emotion or pre-existing religious beliefs of the clients may lead them to suspect demonic activity rather than a ghost.

Indeed, even among some Catholic exorcists, there's a belief that many cases thought to be related to ghosts are actually the result of demons trying to trick people.[1]

1   Rossetti, S. J. (2021). *Diary of an American Exorcist: Demons, Possession, and the Modern-Day Battle against Ancient Evil.* Manchester, NH: Sophia Institute Press, pp. 187-8.

We're not offering any judgments on these beliefs, but if one happens to belong to a religious group that shares them, demonic claims seem more likely.

But beyond the demonic, we've noticed a true smorgasbord of paranormal claims in our private cases. They range from ghosts to demons to invisible jellyfish[2], malevolent clouds, aliens, monsters, giant crickets[3], and so much more. Because of this diversity of claims, we have to be well versed in our lore, and always ready for anything. It also means part of the lifecycle of a private paranormal investigation involves spending a lot of time working with the clients to figure out exactly what their claims are, because it's not just a matter of going to the library and looking them up like we could for a location such as, say, the Elkhorn Lodge.

The cases usually begin with email correspondence. Claims are often vague and we spend some time going back and forth to get more information. We always ask clients to keep a diary of everything that's happening, for our benefit as well as theirs. For them, it helps to organize their thinking and helps them start to sort through which phenomena may or may not be significant before we show up to investigate. For us, it helps pin down the narrative to specific and hopefully testable claims. At the very least, we're looking to find out when the events typically occur so we can investigate at a date and time most likely to yield results.

Strangely, cases often end at this stage. People write to us for a wide variety of reasons and we're always keen to help, but often the correspondence drops off either immediately or after a few messages. Sometimes a message is all that's needed—if we're just analyzing a photograph, we don't need to do much more—but other times the client simply drops out of communication and we can only hope they've gotten whatever help they need.

Reasons people contact us are also varied. Some are experiencing something weird and they just want to know what's going on. We're happy to investigate those and try to answer their questions. Others are terrified, particularly if they think something demonic is tormenting them. They may be writing to us in the hopes of arranging a "cleansing" or a sort of exorcism. We tread carefully in such cases because we don't claim any particular expertise with regard to removing malevolent spirits. At the same time, we're also aware that many other so-called paranormal investigation teams lack our dedication to rigor and can leave people psychologically worse off than before they showed up. So when feasible, we try to stay involved and perform our investigation before referring people to whatever professional may be able to help them, which may include their clergy.

An example might be in order. This won't be one of the cases you'll read about in this book because it's not been completed yet. If it ever is, it may show up in a future volume. We were approached by an individual who claimed to have an evil entity

2   Yes, really. That case will be described in volume 2 of this series.
3   Also a real case. That one is slated for publication in volume 3.

(essentially a demon, though she avoided that word) in her house. Another investigator—an individual whose path we've crossed in the past, so we know how he operates—"confirmed the worst" and then jumped ship, leaving this potential client in a particularly vulnerable mental state.

Our first inclination was (and is) to pursue the investigation and hopefully un-ring the bell the other investigator rung. However, months have passed between communications, though the paranormal claims have grown increasingly threatening between each letter.

Normally, we like to spend a fair amount of time going back and forth to get as much background information as possible before beginning an investigation. At that point, we schedule a meeting on neutral ground (often a coffee shop, to make everyone more comfortable) to discuss the matter in person, and only then do we schedule an on-site investigation. However, given the escalation of the case described above, we've attempted to move forward at an accelerated pace to try to get the client the help she needs (whether that help may be in the form of our investigation, psychological help, consultation with clergy, or whatever else) as quickly as possible.

With all of that in mind, and realizing that especially when it comes to private residences, adaptability is key to a solid investigative protocol, we can divide the typical process into several phases:

1. Initial contact—the client reaches out and we open a case file.

2. Initial discussion—we try to pin down the client's claims as clearly as possible. We recommend keeping a written diary to aid in this process. If relevant and feasible, we recommend setting up recording devices.

3. Background research—based on the client's elaborated claims and diary (if they agree to keep one), we begin conducting whatever background research may be relevant. This may include historical research, folkloric research, or tracing property records, among other things.

4. Initial consultation—perhaps at the same time as we're doing background research, perhaps before, and perhaps after, we schedule an initial consultation with the client. We also want to meet and interview any other people living in the house who may corroborate or contradict the client's story. We may ask the client to fill in a worksheet outlining their experiences, beliefs, any relevant medical history, drug use, major life changes, etc.[4] If they have psychological, medical, or religious professionals with whom they're already working, we may ask for an introduction.

5. Schedule investigation—based on everything determined up to this point, we schedule an investigation at a time that's convenient for us and for the client, but most importantly coincides, if possible, with when the paranormal events typically occur.

---

4   We're not trying to call anyone crazy, but even within paranormal lore, these kinds of factors can be relevant. Importantly, we keep any information we're given in the strictest of confidence.

6. Tour of property—either on an earlier date or, more commonly, at the start of the investigation, we let the client give us a guided tour of the property. We document everything on photo and video and construct a map of the property indicating not only where the paranormal events are supposed to have taken place but also any other items, elements, or appliances that might be relevant.

7. Investigation begins—we ask the client to go about their usual business as they normally would when the paranormal events took place in the past. Meanwhile, we follow much the same protocol as during our public venue investigations and observe silently, taking occasional breaks to take measurements or perform small experiments.

8. Follow-up if necessary—if any questions remain unanswered, we may try to schedule any necessary follow-up investigations or conduct further background research.

9. Issue report—We provide a detailed written report of our findings and also discuss them with the client, taking time to answer any remaining questions. If follow-up is necessary, we schedule it.

10. Provide referrals as necessary—depending on the client's state of mind and our results, we may refer people to other professionals as necessary. These might include clergy or psychological professionals. Again, we always emphasize (truthfully) that we don't refer people to psychologists because we think they're crazy. Rather, we refer them, typically, because they're going through a difficult experience (whether it turns out to be paranormal or not) and often could benefit from professional counsel. We have networks of both psychological and social professionals as well as clergy of a variety of faiths and denominations to whom we can refer people if they don't have trusted professionals of their own already.

As you'll see in the pages to follow, the extent to which we can linearly progress through these stages varies widely. Private cases are fascinating because we never quite know what we're going to get.

In the interest of brevity, we're not going to describe all of these stages in detail during the discussions to follow. Unless absolutely relevant, we'll omit most of the details of our original correspondence and consultations. Simply accept as a given that we did these things before undertaking each investigation.

15

A Military-Owned Residence: The F. E. Warren Air Force Base

Our exploration of the private residence cases begins with a particular case that exists on the boundary between public venues and private residences: a privately occupied but military-owned home on the F. E Warren Air Force base. Because of this unique situation, our research protocol landed somewhere in between what is typical for public venues and private residences. Specifically, we proceeded to investigate the specific residents' claims much as we would a private case but were able to do a bit more historical research as we would for a public case.

Figure 15.1. Artist's rendition of the ghost scaring a child. Artwork: Aaron Bordner.

Though our treatment of the history won't be as complete or detailed as in the public cases, it's worth knowing a little something about the location under investigation. The home we investigated is part of the Francis E. (F.E.) Warren Air Force Base, located about three miles west of Cheyenne, Wyoming. The base is the oldest continually operating Air Force facility, meaning it has plenty of history.

It began its life as Fort D. A. Russell, an army base housing infantry and cavalry. Its location was specifically chosen to allow forces to protect Union Pacific Railroad workers from attacks by what were then called "hostile Indians." It was established in 1867, only one year after Congress formed four regiments of black soldiers (colloquially called "Buffalo Soldiers" at the time). Three of the four—the 9th and 10th Cavalry and the 24th Infantry—served at Fort D. A. Russell.

The connection to Francis Emroy Warren was established in 1903 when one of the residences was assigned to Captain John "Black Jack" Pershing, who would go on to lead American forces in the first World War. Pershing married the daughter of then-U.S. Senator and the first governor of Wyoming Francis Emroy Warren. In 1930, President Herbert Hoover changed the name of the base to Fort Francis E. Warren in honor of the same individual, and it carried on as an Army installation through World War II.

In 1947, the base was transferred to the United States Air Force to serve as a facility for Air Training Command, a duty it fulfilled until Air Training Command transferred to Strategic Air Command in 1958. Its name changed from Fort Francis E. Warren to the Francis E. Warren Air Force Base in 1949. Since 1991, the F.E. Warren Air Force Base has been home to the Twentieth Air Force, which commands the Air Force's entire supply of intercontinental ballistic missile (ICBMs) and serves as one of three strategic missile bases in the United States.

The base is also home to arguably more than its share of paranormal lore. The base personnel were more than happy to share some of their local ghost stories. They even keep a weighty logbook specifically of ghostly occurrences. Whenever someone on the base witnesses something strange, a new entry is added to what fundamentally amounts to an F.E. Warren Air Force Base equivalent of the X-Files. We were allowed to see the book but not to take it away to make copies, so we can't supply a complete report of everything contained therein (nor would we have room in this book to do so even if we could).

The haunted stories of the Warren Air Force base have been featured in several publications. There's even an entire book dedicated to the subject. Individuals stationed at the base have their own Facebook group entirely devoted to sharing the spooky and weird experiences they've had. If we were treating this as one of our public venue investigations, we could easily spend years going through all the lore and meticulously examining every inch of the property. Part of the reason we haven't been able to do that, of course, is that as a military installation that's home to nuclear weapons, the base has some areas to which even such illustrious organizations as Rocky Mountain

Paranormal can't gain access.

To provide a general understanding of the kinds of things claimed to be happening at the base, though, we've selected a small sample of paranormal claims from the area.

By far the most common paranormal sightings involve cavalrymen, apparently from the 19th Century. While many people might struggle to identify them specifically as *19th century* cavalrymen, the history of the base supports this interpretation, and we assume many of the people stationed there are sufficiently versed in military history to correctly identify the era based on the uniforms the ghosts are said to wear. Sometimes witnesses even report communicating with the ghosts. According to one of the communications officers we interviewed, a Staff Sergeant encountered one of these spirits in the 1980s. He greeted the spectral cavalryman with a friendly "good evening."

"Howdy," the spirit replied, then promptly vanished.

Many of the old houses in the older section of the base have their own ghost stories. One involves a service family whose daughter awoke to the sight of a cavalry officer standing in her playroom. Another has even become known as the "ghost house" due to the frequency of paranormal reports. Apparently a captain once lived in the house. When his family returned early, catching him in bed with his mistress, he attempted to escape through a second story window and fell to his death. His ghost has been reported in the house by numerous subsequent occupants over the years. The captain's home office is the room most commonly associated with the tales.

Another tale involving both ghosts and romantic indiscretion is said to have taken place in the 1890s. When an officer was transferred away from the base to another post, he failed to inform his mistress. Lovesick, the woman committed suicide. Her spirit is sometimes seen walking around the upper floor of one of the residences on the base. Tragically romantic ghost stories such as this one are common in both ghost literature and paranormal folklore. Most of us have probably heard at least one story similar to this over the years.

Possibly also related to a romantic story is the case of the woman in black. Warren has its own cemetery, located near the officers' quarters on the base. According to reports from base security officers, a veiled woman dressed in black can often be seen lurking in the graveyard. When approached, she disappears. Though no other information regarding her identity is known, it seems like at least a plausible interpretation is that she's visiting the grave of a lost lover or family member, still grieving the loss even in death.

Yet another residence on base is meant to be haunted by the apparition of an unidentified young girl. On one occasion not long before our investigation (so it was still fresh in the minds of some of the people we interviewed), the home's residents were away on vacation. Despite the house being empty, neighbors saw a young girl staring out of the front window. Worthy of note is that these residents had two sons but no daughters.

Even the general's quarters are not immune to ghostly visitations. Supposedly the house is haunted by a cavalryman and his dog. As we were told, whenever a new general moves into the house and tries to remove a certain picture frame from the attic wall, he's guaranteed to hear eerie noises at night. Such sounds include the barking of a dog, even if the general has no pets. Once the picture frame is returned to its rightful place, the noises and disturbances cease.

Also on the base is the United States Air Force Intercontinental Ballistic Missile Museum. According to the curator, this building is also "very" haunted. Manifestations include the security system frequently disarming itself and the door unlocking and opening itself. That this occurs only at the *museum* rather than a secured military building is a source of significant relief to us. Other manifestations include chandeliers swinging even when no one is near them, lights turning on and off, and other fairly "standard" paranormal claims. An elevator in the building has been rendered unusable because of the frequency with which it traps visitors between floors. Electricians and mechanics have failed to find a cause. Several police canine handlers have said their dogs refuse to enter the building's attic. Bemused museum staff have named the resident ghost "Jeffrey" and say they've learned to live with his antics.

Probably the most famous ghost story on the base is also easily the most disturbing. Reports have been filed by a variety of security officers who regularly drive out to the missile silos to perform security checks as part of their routine rounds. During these outings, some of them say they've seen "Indian braves" riding on horseback on the prairie. While these kinds of apparitions might be expected, a different Native American spirit on the base has a darker backstory. The family camp area (which is near the home at which our own investigation took place, incidentally) was reportedly the scene of a brutal crime in which cavalrymen raped a Native American woman and then bludgeoned her to death. Screams and cries from her ghost have been reported to base security forces on numerous occasions, typically by people who suspected a current (rather than historical) crime was in progress. When patrolmen reach the location to investigate, nothing is ever found.

So clearly, the base has a lot of paranormal lore. Because it's a serious military installation, the people reporting these stories are not mere crazy people. Military service alone does not guarantee one is sane and rational, of course, but the Air Force is not in the habit of letting random loonies loose on a base that houses ICBMs, so the degree of witness credibility in this location is, while not absolute, at least somewhat higher than at many locations.

Our investigation begins at this point. We were contacted by a family who lived on the base and had been experiencing a variety of phenomena they couldn't explain. We will not identify any of the people involved. For purposes of this write-up, we'll call the couple Paul and Jane (not their real names). They also had a son, then eight years of age, who we'll call David. Paul was enlisted in the Air Force and serving on base at the time of our investigation. Jane was similarly employed in a respectable profession.

When we began looking into their claims (which we'll detail in a moment), we mostly followed standard protocols as described earlier for investigating claims at private residences. There were, however, a couple of modifications to the protocol. First, because of the history of the base itself, we were able to do a little bit more background research, some of which we've described above. Second, there were also some practical differences to our protocol. Because the home is on an active military base, we couldn't just enter and leave on a whim and had to involve not only the client family and our team but also Air Force officials in the planning process. Out of respect for the security interests of the United States and USAF, we will not divulge any details of this process or the security protocols required to enter and conduct our investigation on the base.

During our interviews, we developed a positive opinion of our clients and didn't detect any signs of mental disturbance or intent to perpetrate a hoax. Necessarily, this evaluation is largely subjective, but we've developed a pretty good sense for such things over decades of experience and dealing with hundreds of clients ranging from the completely sane and well-adjusted to the completely bonkers (to use the technical terminology). Paul and Jane seemed decidedly on the "well-adjusted" side of that spectrum.

Hoaxes are also something we constantly think about when dealing with private cases, but our experience is that they're quite rare. When they do exist, typically people reach out to newspapers or television stations instead of Rocky Mountain Paranormal, for probably two reasons. First, they want the direct publicity, and second, we have a very strong reputation of exposing hoaxes when they do occur. We saw no evidence that Paul or Jane were trying to pull anything, and they specifically requested anonymity rather than seeking attention for their claims. Most hoaxers can't wait to get in front of the camera. A *skilled* hoaxer (a fairly rare entity in and of itself) might feign disinterest in publicity specifically to avoid detection, but we're convinced that's not what was happening in this case.

Furthermore, both adults in the family held highly respectable positions and were approved by the Air Force to work on a base that houses nuclear weapons, so we assume it's the opinion of the United States government that these are sane people. The unanimous evaluation of Rocky Mountain Paranormal's team concurred with that (presumed) assessment.

We emphasize the credibility of the clients in this case even more than in some other cases because what they told us they'd been experiencing is a bit more dramatic than a lot of the claims we get. When people hear unexplained noises, it's easy to dismiss as simply house noises for which they haven't yet found explanation. Paul and Jane, though, had much more tangible claims.

The events began in their house shortly after they moved into the two-story semi-detached residence a short time before our investigation (we do know the exact year, but are withholding that date to remove one more potentially identifying piece of information from the record). The building is owned by the military but is a single-family residence.

About a month after moving in, Jane experienced the first allegedly paranormal phenomenon. While walking down the stairs, she encountered a Native American male who appeared to be between the ages of 40 and 50 years. She described him as having a "jowly" face which looked "very disapproving, like he had caught me with my hand in the cookie jar." His hair was long and thinning, oily black and tied back. He wore a t-shirt. Jane couldn't recall his pants or footwear. She screamed and ran into the bathroom, locking the door behind her. While inside, she said she heard a young female voice crying and screaming. Though we can't make a definitive connection between the two, we do note that not far from the house is the family camp area said to be the site of a rape and murder and which has subsequently been known for ghostly manifestations of a young woman screaming and crying.

The next month, upon returning home from duty, Paul saw a dark human silhouette or shadow in the front window. Upon entering the house, he discovered no one was present. He did see a smaller black shadow which he described as the size of a dog moving down the hallway at high speed. The shadow made a sharp left turn into the kitchen and then disappeared through the closed and locked back door. Importantly, no pets were in the house at the time. Official records can confirm that police were called to the house on suspicion of a potential home intrusion. They swept the house but found nothing. After looking around, the sergeant told Paul these kinds of events were "not an unusual situation" on the base.

Young David did not escape having his own experiences, either. A couple months later, in November, he awoke to the sight of a Caucasian male of indeterminate age standing behind his chair in the bedroom. When the family reached this part of their story during our interview, we immediately separated David from his parents to question them independently.[1] David provided a precise description of the man: he had gray hair and a long beard, with a modern-style blue work shirt and denim trousers. David's interpretation was that he was a "nice guy; he was kinda smiling." At the end of the encounter, the man walked across the room to the window and disappeared. Questioned separately, Paul and Jane confirmed David had told them substantially the same story, with no major alterations.

Since this disturbance, and until the time of our investigation, David refused to be separated from his parents. He took to sitting outside the bathroom door with his mother inside and often demanded to sleep with his parents.

The child wasn't alone. As events continued to deteriorate, Paul and Jane took

---

1   Interviewing witnesses separately is always a good idea. When children are involved, it becomes even more important, because children are highly susceptible to strong but often unconscious influence from parents or other authority figures.

to traveling in pairs within their own home, each refusing to go downstairs for water alone at night.

During the six weeks before our investigation, the family regularly heard footsteps on the staircase at night. Cupboard doors and drawers in the kitchen would open and close by themselves and could be heard slamming at night when all the family members were in bed.

And then, the strangest and most frightening thing at all.

One morning in December, Paul and Jane went downstairs in the morning and found an unknown substance smeared all over their refrigerator.

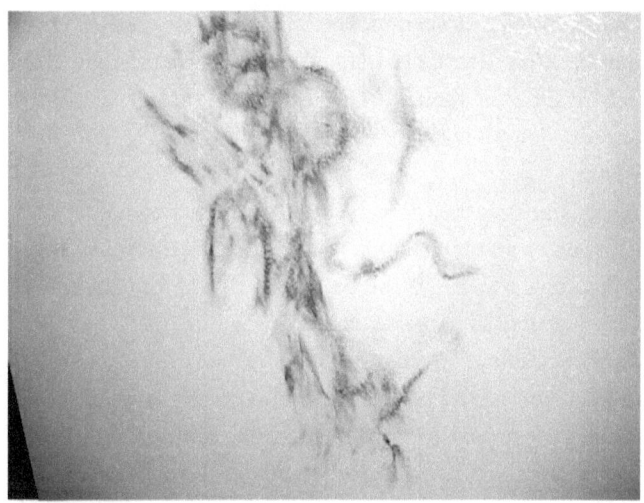

Figure 15.2. Unknown substance on refrigerator door. Though presented here in black and white, the substance consists of red and brown colors. Photo provided by the client.

A photograph of the mysterious substance, taken by either Paul or Jane, was provided to us prior to our investigation. The substance consisted of brown and red colors. To Paul's eye, it looked like it spelled the words "go now." Primed to see that specific message, we can easily make it out in the photograph. We're not sure whether we would independently be able to decipher it as spelling out the same words. Regardless, whether or not Paul's interpretation of the message is correct, there absolutely was this unknown messy substance on the fridge.

The family described the substance as having a "sweet smell" and spent the morning going through all the food in the cupboards to try to figure out what it might be. Paul claimed to have tried to match the substance's appearance using peanut butter and jam (our own first guess) as well as other substances, but he couldn't find anything that he thought looked remotely like whatever was smeared on the refrigerator door.

They never figured out what it was. They said they kept a small sample for future analysis, but if they ever did that analysis, they never told us the results. We would have loved to take a sample ourselves, for analysis either within our own in-house laboratory or at an independent professional lab. However, we were unable to get access to a sample, so we also don't know what the material was.

Though we'd certainly like to know, it almost doesn't matter. Even if the material did turn out to be something as mundane as peanut butter and jelly (as one of our members remarked upon first seeing the photograph: "it looks like someone stuck a cherry bomb in a PB & J sandwich, but I wouldn't bet the farm on it"), it's still a mystery as to how it came to be on the fridge. If we take the family as trustworthy (as we're inclined to do) and they claim not to have made a mess of their own fridge, *something* unusual has to be afoot, whether paranormal or otherwise. Perhaps the most mundane explanation might be that their child made the mess and later lied about it, but David was an older child at the time than we'd expect would make such a mess and struck us as honest during our interview.

We simply don't know.

Shortly before our investigation, Paul and Jane acquired two dogs. They told us those dogs had begun to stare into thin air and chase an invisible "something" around the bedroom. We didn't observe these behaviors during our investigation, so we can only speculate. One potential naturalistic explanation is the "zoomies," known to animal behaviorists and veterinarians in more technical terms as "frenetic random activity periods." It's a well-documented phenomenon that many animals (including dogs) will engage in spontaneous running sessions, sometimes running in circles or up and down the hall. This is considered a normal and healthy animal behavior, particularly characteristic of puppies or young dogs (as this family's dogs were) in which they release much of their built-up energy all at once. We're not necessarily saying "zoomies" *is* the explanation here, but it is entirely possible that people who are in a state of mind to look for paranormal phenomena due to their other experiences may have ascribed a supernatural origin to a natural behavior.

Cold spots were frequently felt throughout the house. Paul described the atmosphere as feeling "like a meat locker" on several distinct occasions.

Cold spots are often reported in supposedly haunted locations. One theory meant to explain these is that spirits draw energy from their environment in order to manifest, and this creates the cold spot. We don't particularly like that explanation. Going back to Physics 101, recall that any exchange of energy produces heat as a byproduct. That's not to say it's impossible to create "cold spots." Your refrigerator is a good example of a cold spot. However, if you ever stick a thermometer behind your fridge, you'll quickly realize that refrigerators generate a lot of heat in order to cool down their interiors. So while it's possible to create a cold spot (even hypothetically by paranormal means), the laws of physics tell us we should also be looking for a coincident hot spot.

After we started explaining this idea in some of our lectures, we noticed that some other paranormal groups have started talking about "hot spots" in supposedly haunted locations. We're not sure if that's because of our lectures, but if not, it's a bit of a coincidence.

Either way, when we're on an investigation, we're always happy to look for *any* thermal anomaly, whether hot or cold. But we've noticed that some (not all) other groups use their thermometers incorrectly or misinterpret the results, so we urge caution when looking into these claims just as we do with all claims of paranormal phenomena.

Furthermore, objects began to disappear around the house. Sometimes they disappeared permanently (some cookware) and other times they would reappear in other parts of the house (a hammer). On one occasion, Paul came home to find a candle burning, but Jane insisted she hadn't lit the candle before leaving the house.

At the end of our interviews with the family, we also took note of the common paranormal claims they did *not* include in their narrative. They didn't report any rapping or knocking on the walls, olfactory phenomena, or strange voices within the house. None of the family members reported nightmares or strange dreams. Absence of these phenomena doesn't mean anything conclusively, but we do note that all of these are commonly reported claims in similar cases, so it's noteworthy that this case is a little bit different from the rest of the pack.

Our investigative team arrived on-site at about 4:00 p.m. on the day of our investigation, after passing through the requisite military base security checks and we started off by conducting our interviews and then took a brief tour of the house. We noted that it was an exceptionally windy day, with gusts reaching up to 50 miles per hour according to base weather reports. Weather conditions often seem to be less than ideal during our investigations for some reason. Nevertheless, we moved forward with the investigation and after having dinner, we started the investigation proper at about 7:00 p.m.

In terms of monitoring equipment, we wired the entire upper floor of the house with microphones, video cameras (including infrared night-vision) and thermometers. We placed control objects at the location of the apparition sighting in David's bedroom. Recall that control objects are items placed in a location whose precise location and orientation are documented before, after, and during the investigation to detect even small movements or changes.

Observers took posts in David's bedroom, on the stairs, and downstairs.

We took baseline EMF readings throughout the house and found levels to be normal. The only EMF "spikes" we detected were the expected and usual ones near the fridge, microwave, lighting, and outlets. Levels on the EMF readings throughout the evening remained normal and steady across the board.

We did detect a minor temperature anomaly whose significance is uncertain. Temperature readings upstairs remained 4-6 degrees (Fahrenheit) colder than downstairs throughout the evening (upstairs temperatures ranged from 64-66 degrees and downstairs from 70-72 degrees). Paul and Jane told us this was the opposite of what they normally experienced. The house was typically extremely hot and the lower floor cooler.

In houses in general, neither is universally true. Depending on the design of the house and the heating/cooling systems, sometimes upstairs rooms are warmer and sometimes colder than downstairs ones. The conditions described by Paul and Jane as normal are consistent with a well known phenomenon in which warmer air rises to the upper floor of a house, displacing cooler air to the lower floors. Typically, though, the temperature differentials between upper and lower floors are consistent within a house. It's unusual that the upstairs should be cooler during our investigation when it's usually warmer.

One possible explanation was the uncharacteristically high wind on the evening in question. Depending, again, on the house's construction and insulation, the wind may have accelerated heat loss through the upper floor of the house faster than on the lower floor.

That explanation may coincide with one strange event that occurred during the investigation. At about 10:10 p.m., we noticed a window in the master bedroom had opened seemingly by itself. It was closed when we had swept the room for EMF readings earlier in the evening but was open when we checked the room again later. After we closed it this time, it remained shut throughout the remainder of the evening.

Perhaps the open window could partially explain the unusual temperature differential. Especially if the window wasn't entirely sealed, heat may have escaped through the crack. Similarly, if it wasn't properly latched, the wind could have caused it to open. However, if we assume the window was closed and latched, then we have no explanation for how it could have opened between our sweeps of the room.

We did hear some unusual tapping sounds in David's bedroom during the night. Wind again provides a plausible natural explanation, but to be certain we'd have to follow up on a less-windy day.

None of the control objects moved throughout the investigation, and we didn't experience anything else unusual during our stay, and we packed up and went home shortly before dawn.

We were somewhat disappointed that we couldn't document more during our investigation, particularly given the level of disturbances reported. Paul and Jane did tell us the disturbances worsened when they experienced periods of domestic upheaval (even arguing over "something stupid"), which was not the case during our investigation. They also said things worsened during the full moon.

On those notes, we have a few thoughts. First, the so-called "full moon effect," or the popular belief that people behave strangely during the full moon, has been solidly debunked by other researchers' statistical analyses.[2] Whether paranormal phenomena

2   Cambell, D. E., & Beets, J. L. (1978). Lunacy and the moon. *Psychological Bulletin, 85*(5): 1123-1129.

ebb and flow according to lunar cycles, to our knowledge, has not yet been the subject of sufficient mathematical analysis. Unfortunately, unlike criminal, medical, or psychiatric events, there is no official dataset representing paranormal claims. Whether extant data on the matter are sufficient for statistical analysis is an open question and one we hope to pursue in the future.

With regard to periods of domestic trouble coinciding with paranormal phenomena, we're not particularly surprised. There are two potential explanations—one paranormal, and one psychological. In much paranormal lore, experiences coincide with periods of substantial turmoil. Some people believe negative emotions invite sinister entities. Others believe spirits may appear to guide people through periods of difficulty. Either way, there is substantial paranormal lore to justify consideration of the idea that paranormal phenomena could coincide with marital disputes. On the other hand, a justifiable skeptical explanation is that hallucinations, delusions, or other psychological effects that could result in erroneous belief in a paranormal experience might also be heightened during periods of psychological turmoil.

Unfortunately, our investigation did not occur during either a full moon nor a time when the couple were having a dispute, so all we can do is speculate.

When we left, we hoped to return to the base shortly for a follow-up investigation (preferably during a full moon on a less windy day). However, Paul and Jane transferred to another base before we were able to schedule such a follow-up investigation, leaving us, as so often happens in this line of work, with far more questions than answers, both about the particular house in question and about the reported hauntings of the F. E. Warren Air Force Base in general.

References and Further Reading

Cambell, D. E., & Beets, J. L. (1978). Lunacy and the moon. *Psychological Bulletin, 85*(5): 1123-1129.

Iosif, A. & Ballon, B. (2005). Bad Moon Rising: the persistent belief in lunar connections to madness. *Canadian Medical Association Journal, 173*(12): 1498-1500.

Pope, J. (2014). *Haunted Warren Air Force Base*. Charleston, SC: Haunted America, The History Press.

Rotton, J., & Kelley, I. W. (1985). Much ado about the full moon: A meta-analysis of lunar-lunacy research. *Psychological Bulletin, 97*(2): 286-306.

Rotton, J., & Kelley, I. W. (1985). Much ado about the full moon: A meta-analysis of lunar-lunacy research. *Psychological Bulletin, 97*(2): 286-306.

Iosif, A. & Ballon, B. (2005). Bad Moon Rising: the persistent belief in lunar connections to madness. *Canadian Medical Association Journal, 173*(12): 1498-1500.

16
Smudging the Demon

Not all of our cases involve lengthy and detailed investigations. This is an example of one of the briefer ones. It took place at the home of a single mother and her son in Westminster, Colorado. Despite the limited detail we can offer, there are some important lessons regarding our protocols and some important ethical considerations surrounding this case.

Figure 16.1. Artist's rendition of smudging the demon. Artwork: Aaron Bordner.

This case involved a family consisting of a single mother in her twenties and her son, aged eight years. In order to protect their identities, we'll call the mother Alice and the son Billy.

They'd lived in their home for several years without any sign of a paranormal incident, though they described themselves as a very "spiritual" family. Some time before we got involved, they began experiencing some unusual phenomena.

Repeatedly over several evenings, Billy saw a man in his bedroom. By the time we got involved, he refused to even enter the room. Around the same time, Alice began seeing an apparition of a woman walking across the living room and entering the kitchen. Both of these phenomena were recurring and neither Alice nor Billy could identify either of the people they'd seen. Around the same time, they struggled with issues of items being knocked from shelves throughout the home.

Clearly, they were terrified. They didn't know exactly what was happening, and it was interfering with their lives. Alice reached out to us for help, which we were keen to provide. One of the most important aspects of our work is the opportunity to help people get their regular lives back. When someone is so confused or afraid that they can't even enter their own bedroom—what should be one's personal sanctuary—something obviously needs to be done.

Our typical mode of operation is to undertake our usual investigative process. People who reach out to us may not have a scientific or forensic investigation in mind—they might be looking for something more like an exorcism or spiritual guidance—but we've found that it's necessary first to determine what's happening and then to determine the correct course of action. Often, as you'll read in some of the cases to follow, we're able to identify natural causes for whatever disturbance had been plaguing the client. When that occurs, it's a great relief to all involved because they now have the explanation and no longer need to be afraid of their own house. You would be amazed by the psychological difference that can make in someone's life. On the other hand, if we were to determine that the house actually was plagued by some kind of evil entity, the investigative work still wouldn't be in vain. It could provide the critical insights necessary to formulate a plan of further action.

When we arrived to interview Alice, she was alone in the home. She'd sent Billy to a local park to play with a friend so he didn't have to be involved in the proceedings. After a tour of the house, focusing specifically on the bedroom, hallway, and kitchen, we discussed next steps with Alice. We explained how our investigations usually work (repeating some information that had already been discussed in preliminary correspondence) and shared our reasons for proceeding the way we do. Alice, though, wouldn't have any of it. She had no interest in scientific investigation and insisted only that someone bless her house.

This is a point of some delicacy. Many paranormal investigators on the skeptical side would never dream of doing such a thing. Even if they're skeptical in the true sense of the word and not merely cynical disbelievers, they have a point. Assuming the paranormal belief at least *might* be delusional, then undertaking a blessing, exorcism, or any other spiritual practice could easily reinforce that delusional belief. There's certainly an ethical case to be made against such actions.

Along the same lines, though we have developed some expertise in both paranormal lore, investigative techniques, and several religions during the course of our work, we are not clergy and do not consider ourselves qualified to perform these kinds of rituals. Even if we were, the specific rituals or doctrines with which we might be familiar might not align with the client's own. That, in and of itself, raises an interesting ethical question. We maintain that there actually are right and wrong answers regarding such things. Ghosts, demons, gods, or any other entities either exist or do not exist, and that's an objective reality. If we could prove the matter one way or the other, we would feel an ethical obligation to share the truth publicly. Absent that kind of proof, however, we make it an important point of our professional ethics to conduct our work within the context of the client's own religious beliefs. We're looking for answers, not trying to force our own beliefs or philosophy upon anyone who may be in a vulnerable state.

So there are plenty of good reasons not to get involved with something like a blessing of a client's house. But there's also a counterargument.

Clients who reach out to us are often at the very end of their tether. We've even had clients so terrified of the paranormal claims that they've been prepared to walk away from their mortgages even though it would spell financial ruin. If we're in a position to help people through such a situation in such a way that they can ameliorate their fears and get their lives back in order, we feel an ethical obligation to do so. We don't believe in turning people away who desperately need assistance and who have come to us seeking the same.

In some cases, the assistance we could provide might involve referring people to psychological professionals. In other cases, we may choose to work with the client's own clergy or refer them to clergy members aligned with their own religious beliefs.

For Alice and Billy, we determined this was the best course of action. Through our discussions, we had determined Alice was of a Pagan belief system, so we tracked down a practitioner and arranged for a blessing of the house, which involved "smudging," a practice of cleansing a location of negative "energy" and replacing it with positive through the burning of white sage. Smudging is a practice with origins in the spiritual practices of certain indigenous peoples of North America, though it has since been incorporated into a variety of other New Age or spiritual belief systems and practices. Because this practice is not part of our own training or philosophy, we found a reputable practitioner to perform the ritual, in which we would not participate directly. Instead, we were present to observe, document, and to provide any further assistance necessary.

A mutual agreement was reached between Alice and our team that Billy should not be made aware of the ritual. For Alice's part, she didn't want her son to be afraid of what was going on in the house. On the other hand, for our part, we were interested in seeing whether the paranormal allegations would cease even if he didn't know that a ritual had been performed.

After the ritual was complete, we departed. A few weeks later, we reached out to Alice to inquire as to the current status of the case. She said everything was back to normal. Several years later, she reached out to us and thanked us again for our assistance.

This result leaves us with more questions than answers, but there seem to be two major competing hypotheses:

1) There actually was a paranormal presence in the house and the blessing/smudging ritual succeeded in either exorcising or pacifying it.

2) The paranormal claims stemmed from Alice's own psychological belief in the existence of the entity, and Billy picked up on her distress. Once she was convinced it was gone, she returned to a normal psychological state and everything went back to normal.

We lack the evidence to make any definitive determination, but we've grown quite accustomed to this dichotomy. Throughout this series, you'll see several cases in which a spiritual cure can be seen as successful even though we remain unable to determine whether the mechanism was supernatural or psychological. We may have our suspicions, but the end result is largely the same. Indeed, even the practice of dynamic psychiatry historically grew out of the tradition of exorcism.[1] The world is a strange place, and the human mind is arguably even stranger.

---

1   Ellenberger, H. F. (1970). *The Discovery of the Unconscious: The History and Evolution of Dynamic Psychiatry*. New York: Basic Books.

17
The Cancer Patient

People contact Rocky Mountain Paranormal (or any paranormal research group) for a variety of reasons. Sometimes they want an investigation just to figure out what's going on. Other times, as in the previous case, they're looking for a blessing or exorcism. Still other times, their reasons are something entirely different. Once we're able to figure out what the real reasons are, we're in a better position to assist, as is illustrated by this case which took place at an apartment in Boulder, Colorado.

Figure 17.1. Artist's rendition of the cancer patient. Artwork: Aaron Bordner.

When we were called to the home of Eileen (not her real name), we expected just another run of the mill paranormal investigation (if there is such a thing; the longer we do this, the more we become convinced that every case truly is unique). We quickly learned that the case was actually quite different from what we initially expected.

The home was a small but tasteful apartment in Boulder, Colorado. Eileen herself was a single woman who appeared to be in her forties and was living alone.

She called us complaining of a variety of fairly common paranormal phenomena, which further made us think it was going to be a fairly common case. Over the preceding weeks, she'd been hearing voices and footsteps throughout the apartment and seen people walking around her home late at night.

With regard to the former, we suspected the culprit could have simply been the apartment setting. Sound can bleed through walls in many apartment buildings, and if a noisier set of new neighbors had recently moved in, their own comings and goings might account for some of the strange noises. Of course that's not to say we were convinced that was the case—as we have repeated many times and probably will many more, we're committed to neutrality when we begin investigating a case—but we noted it as one possible natural explanation to look into.

Seeing apparitions in the apartment, though, is a bit harder to explain. If one discounts the possibility of either a home intruder or a hallucination, it becomes a lot more challenging to think of hypotheses to investigate, so we were keen to get started.

We arrived at the apartment in the late afternoon and sat down with Eileen to discuss her claims and begin formulating a plan for our investigation. Our initial impression of Eileen was of a fundamentally sane and level-headed person. Once again, this made us think it was going to be a fairly standard case.

As the story carried on, however, it began to grow more extreme. Whenever one of our people would interject a potential explanation for one of her claims, Eileen would add new elements to the story. Initial claims were what we'd call run of the mill, commonly reported paranormal allegations. Noises around the apartment, shapes seen out of the corner of the eye, things like this. As we attempted hypothetical explanations for each phenomenon, Eileen would say something like, "Sure, but can you explain…" and follow-up with increasingly bizarre claims. Floating objects, doors opening and closing of their own accord, unexplained menacing voices, and even more occurrences found their way into the story.

By the end of the story, it no longer bore much similarity to the kind of classic case we thought it was. We were admittedly a bit confused for a while. This kind of growing story is atypical of the kind of sane person Eileen seemed to be. And if the story was as extreme all along as it was by the end of the story, why hadn't she told us the whole story when she first reached out to us?

When we reached the stage of our interview in which we ask about major life changes, we finally got our answer. Eileen had recently been diagnosed with terminal cancer.

Living—and now dying—alone, Eileen had begun (unconsciously, at least at first) to invent stories to attract visitors to her house. She began with the local emergency services, but when they stopped responding, she turned to paranormal investigators, and our website was the one she happened to land on. Thank God! Though there are other reputable paranormal investigation groups out there, there are also plenty of fly by night operations who would *not* have handled this case correctly. All she wanted was someone to talk to about her life and her fear of death.

Immediately, our team members exchanged a look. Without exchanging words, we knew exactly what we needed to do. We gave up on any further discussion of paranormal investigation. Instead, we spent the night sitting in Eileen's apartment, listening to her stories and occasionally interjecting a few of our own.

It's amazing how little human kindness people need to get through their lives and it's tragic that some people have to resort to calling paranormal investigators just to get it. But we're honored to have been able to do so. As we said in this chapter's introduction, people call paranormal investigators for all kinds of reasons, and once we know what the reasons are, we can respond appropriately. It's also a good lesson for all of us in this line of work: put aside your preconceived notions of what any given case might be about and actually *listen* to your clients. Given the opportunity, they'll usually tell you what they need.

No further ghostly phenomena occurred for the remainder of Eileen's life.

18
Native American Spirits

Something we've noticed over the years is that a lot of the ghost stories we encounter, particularly in the American west, involve Native American spirits. There are many reasons for this, and a lot of times it seems to simply be a label placed on an anomalous experience absent any better explanation. Occasionally, however, there seems to be a more direct connection between the paranormal claim and the Native Americans.

This case took place on a Native American reservation in southern Colorado. We can't say for certain whether the spirits alleged in the claim were Native American themselves, but it does represent one of the cases that most directly involves the Native American communities. It's also a particularly terrifying collection of paranormal claims.

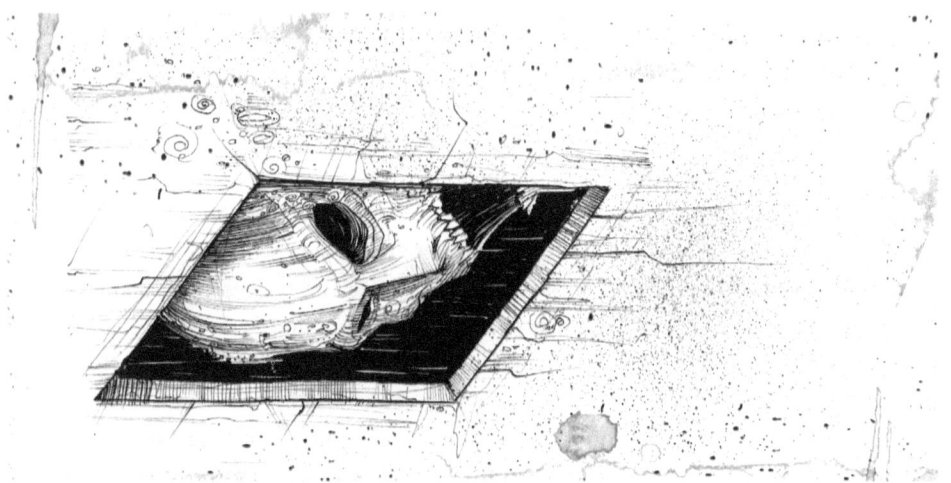

Figure 18.1. Artist's rendition of the ghost at the attic door. Artwork: Aaron Bordner.

As we mentioned in the introduction, we see a lot of cases involving allegations of Native American spirits in our work. Over the years, we've developed some theories as to why that might be. Of course one possible interpretation, and one which we can never entirely discount, is that there simply *are* Native American spirits wandering the American west. Still, we've given some thought to other possible reasons they may be disproportionately represented in western American paranormal lore.

One possibility is that many Americans seem to have developed a romanticized view of Native American culture without ever fully understanding it. The stereotype of Native Americans as particularly spiritual people is well-known. Reality is more complicated. Spiritual beliefs among the various indigenous tribes in North America are far from uniform, and the ideas many Americans have of their spiritual beliefs simply aren't correct. But because the stereotype is so prevalent, it's not beyond the realm of plausibility to think that this stereotypical spiritual orientation could survive death and lead to a disproportionate number of ghosts.

There's also the popular legend of the "vengeful spirit," featured prominently in fiction of ghosts and the paranormal. Given the often-harsh history of mistreatment of the Native American peoples over the last few centuries (which is undeniable, though itself often oversimplified and misunderstood), history and paranormal lore could be dovetailing here in some unconscious manner.

We also can't overlook the fact that a lot of stories from horror fiction involve ghosts haunting Native American burial grounds. We've seen some of those stories already, and there are even more of them to come in later volumes of this series. The idea seems to be that desecration of sacred Native American burial ground can produce a haunting. It's rather difficult to verify which cases this might apply to, though. Not all of the Native American tribes had the same religious or burial practices (the view of the Native Americans as a homogeneous community is simply false), and records don't exist for large numbers of burial grounds. Those grounds for which records do exist don't seem on a cursory examination to have any more allegations of paranormal activity than any other areas. It's further complicated by the fact that, at some point in history, probably the entire country—nay, the entire world—has been burial ground for some civilization or other. Anthropologists have long told us that the first Native Americans arrived on the continent after crossing a land bridge from Asia and Siberia some 13,000 years ago. More recent evidence has suggested that the earliest communities to people North America might have been here as long as 20,000 years before that.[1] Either way, people have been here a long time, and it's likely that someone's been buried on just about every plot of land in the country at some point in history.

Regardless of the stereotypes surrounding Native American beliefs, spiritual practices, or ghosts, we have quite an interest in the reality of the situation. We've

1   Becerra-Valdivia, L., & Higham, T. (2020). The timing and effect of the earliest human arrivals in North America. *Nature, 584:* 93-97.

documented diverse paranormal claims and legends from all different kinds of people, including some of our Native American colleagues, and have begun to develop an understanding both of traditional Native American belief systems as well as how those have developed in more recent years through a certain religious cross-pollination with European religious ideas.

Several years ago, we were called to a home located on a Native American reservation in southern Colorado. The resident family consisted of a mother and father who we'll call Steven and Rachel, both in their late twenties, and their son, who we'll call Jacob, aged seven years. They were completely terrified of the events they'd been experiencing in their home. Truth be told, not a whole lot scares us. Both by nature and by experience, most of our crew are the kinds of people who "ain't 'fraid of no ghost" and are in the habit of running toward instead of away from potentially threatening events. But even we have to admit that the stories they told us, if we'd experienced all of them, would have given us a Grade-A case of the Jeebies.

The ghost stories didn't begin with Steven, Rachel, and Jacob. Several generations of the same family had occupied the house and some of the stories had been shared through the generations. Most prominent among these is the story that took place in the child's (Jacob's, at the time of our investigation) bedroom. Within that room is the door to access the attic. The story goes that late at night, a ghost would scare the children by peeking from the attic access door. Multiple generations of children told fundamentally the same story, describing the face as thin and dark, almost like a shadow, but still with some degree of substance to it.

Steven and Rachel said when they were in the living room watching television together, Jacob would be in his bedroom. On one occasion, they heard a conversation from the child's room. Not only that, but they specifically heard two distinct voices. Wondering who else was in the room, they sprung up to investigate. Upon entering the room, though, they found only Jacob, fast asleep on the floor. Especially since they said they found Jacob on the floor, one skeptical take is that he was simply awake and playing past his bedtime and quickly faked being asleep on the floor when his parents arrived. This, however plausible, would not account for the two distinct voices the parents reported hearing.

On another occasion, in the middle of the night, the two adults heard the sounds of Jacob's toys—beeping robots, sirens on toy cars, and the like. Figuring their son was awake and playing well past his bedtime, they went to the room to investigate (and perhaps to gently reprimand), but once again found the lad fast asleep in his crib (this was some years before we began our investigation and he was still sleeping in a crib at the time).

Yet another story was related by Rachel, who said she and Steven were in bed and both asleep. When Rachel awoke in the middle of the night, she saw a woman's figure at the foot of the bed but could not move, scream, or ask her husband for help. After a few seconds, the figure floated above the bed and eventually disappeared.

Though we weren't present at the time to investigate properly and can therefore only speculate, this story does sound reminiscent of a known psychological phenomenon known as sleep paralysis. To explain a complicated phenomenon in brief, when an individual falls asleep, the brain does not completely shut down. Instead it produces a series of dreams. Because the body is still connected to the brain, it would be possible for an individual to act out his or her dreams in physical reality, which obviously poses any number of risks. Somnambulism, or sleep walking, is one variety of this, and patients with that condition can indeed get themselves unconsciously into dangerous situations. To avoid this problem, the brain normally paralyzes the body during sleep. Sometimes, though, the brain begins to wake up a little bit sooner than the body does. When this occurs, the brain, in a condition somewhere between sleep and wakefulness, projects hallucinations into the waking world, but the individual experiencing them is still paralyzed and unable to respond. The nature of the hallucinations produced seem to be culturally determined. Many scholars believe medieval reports of demonic assaults by incubi or succubi were actually episodes of sleep paralysis. On the other hand, scholars also attribute many modern cases of alien abduction to a different sociopsychological interpretation of the same psychophysiological phenomenon.

While we have to remain open to the possibility of a paranormal interpretation, Rachel's story with regard to this particular episode so closely matches the stories told by any number of sleep paralysis patients that we consider it a highly probable explanation.

Steven, on the other hand, told yet another story. One night he awoke in the middle of the night and saw a young Native American girl standing between his bed and his son's room. The ghostly visitor just stared at him. He turned to wake Rachel up but she wouldn't awaken. When he turned back, the girl was gone.

Family lore also tells of a ghost who would sit outside the home and cry. Though different in the details, we were reminded of the Mexican and Mexican-American legend of La Llorona, literally The Weeping Woman, who according to the folklore is the spirit of a woman who drowned her children after discovering her husband's infidelity and is now cursed to walk the earth, constantly crying, until she finds her lost children.

Rocky Mountain Paranormal was initially contacted by Rachel, who was concerned for the safety of Jacob and wanted us to determine whether there was any kind of imminent threat. When we arrived, only Rachel and Jacob were present—Steven was still at work at the time. Rachel sent Jacob to another room so we could discuss the haunting events openly without scaring the child.

During the initial interview with Rachel, Steven returned home, driving a semi. He entered the house appearing in dress and mannerism to be the stereotypical "trucker" type, and some of our crew wondered what he would make of the presence of paranormal investigators in his house. It's admittedly a stereotype and certainly not universally true, but it's been our experience that often one member of a couple (usually the man) disbelieves in the paranormal claims or disapproves of the paranormal investigators while the other member (usually but not always the woman) wants to pursue the matter. This case did not follow that stereotype at all. As soon as we introduced ourselves to Steven, he immediately sat down and started sharing his own ghost stories. As he spoke, he became visibly shaken and began chain smoking. Indeed, he appeared to be the most frightened member of the entire family, to the extent that he'd begun sleeping with a firearm under his pillow.

One rare benefit we had during this investigation that we almost never enjoy on our private cases was that we were able to interview not only the current occupants but previous residents of the home. Because it had been in the same family for generations, we were able to verify that previous occupants had also experienced unexplained events, and many of them were still afraid of the property.

We were also able to obtain the property records and determined that before the home had been constructed, the property had just been an empty field (that is, there was no prior structure on the land). We weren't, therefore, able to find any information regarding crimes or other historic events on the property that might be relevant to the paranormal lore. It seems like just another house, albeit one that has more than its share of ghost stories.

After the interview and tour of the home, we began our investigation and monitored the home for the remainder of the night using our usual techniques (including video and audio recording, EMF sweeps, temperature monitoring, etc.). Unfortunately, nothing unusual happened during the entire investigation, so we weren't able to provide many answers.

When we completed our report, we shared some of our hypotheses with Steven and Rachel including our suspicion that at least some of the phenomena could be due to a sleep paralysis episode, but we also had to admit that we simply didn't witness most of the kinds of things they'd been seeing. We did explain that we've never come across a documented case in which paranormal activity has caused an individual any physical harm, so we tried to put their minds at ease a bit that they didn't need to worry for their family's safety. Alas, that's about all we could do for them.

The unfortunate reality in paranormal investigation is that occasionally we get a case file that's all build-up and no payoff. This is such a case. All the stories the family told us were beyond intriguing, but the investigation itself simply didn't produce any results. Field science in general is much harder than laboratory science, and paranormal research in particular can be an endless source of such frustrations.

We told the family to contact us if anything else happened and we'd be more than happy to follow up, but they haven't reached out again.

19
The Portal Ghost

We'll take on almost any case a potential client can bring to us, as long as there's at least some chance we can think of a method of investigating, documenting, researching, or otherwise helping the clients uncover the truth and move on with their lives. That said, we prefer to be the first team someone calls. If someone's already worked with another group, we're still more than happy to work with them, but our experience has been that not all paranormal research groups are equally created, and some of the other groups have given clients false impressions.

This case involves a family with numerous paranormal claims, but since we weren't the first group contacted, much of the story had already been contaminated by prior paranormal groups, so it was difficult to separate which parts of the story came from which source. Nevertheless, we successfully completed an investigation of this Littleton, Colorado home, and our results were somewhat different from the other teams'.

Figure 19.1. Artist's rendition of the portal ghost. Artwork: Aaron Bordner.

At the time this case occurred, we were a member of the "TAPS Family" of paranormal investigation groups and received a number of referrals from the TAPS people for cases in our region. It was a good arrangement. We were able to take on cases when TAPS couldn't justify the travel expenses to this part of the country and in exchange they provided us with some interesting work. Unfortunately, sometimes we weren't the first group to which the clients were referred, and sometimes that made our investigations more difficult than they needed to be. This was such a case.

The family involved in this case consisted of a mother (middle aged), an adult daughter, and a preteen granddaughter. For this discussion, we'll call them Mary, Judith, and Kelly, respectively. Mary's husband was also a resident of the home but was not involved in any of the paranormal claims or investigations; we never met him or had any communication with him during the entire process. No other individuals lived in the home (located in Littleton, Colorado) at the time of our investigation. The house itself was a split-level home featuring three bedrooms, one full bath, one fixed bath, one three-fourths bath, one fireplace, with an attached garage and enclosed porch. During our background research, we were able to determine that the house was built in the 1970s and had previously been sold several times in the 1970s and 1990s (we do know the exact dates; we're not providing them so as to maintain the clients' anonymity). Previous occupants never reported any paranormal activity as far as our research was able to determine.

The story begins before our own involvement in the case, and the family reported quite a list of paranormal goings-on. We'll begin by listing the various experiences they relayed to us.

Just after moving in, they said they smelled cigarette smoke at the base of the main stairs leading to the second floor. None of the family members were smokers and they said this couldn't have been residual smoke in the house from a previous occupant.

Several times, they also heard footsteps on those main stairs. Additionally, they saw "shadows" moving around the entire home. Multiple people, including the residents and some of their visitors, reported being bumped by something or someone that wasn't really there.

During one Halloween season, several unexplained things occurred. The family had decorated for the season with lights on the front of the home. At the end of the evening, they would unplug those lights. However, one morning, they'd been mysteriously plugged back in while the family were asleep. Judith and Kelly further reported they'd both heard a loud party coming from the kitchen during the same Halloween season even though no one else was in the home.

The house's basement had been converted into a playroom for Kelly. She reported that she'd been seeing shadow people in that room. "Shadow people" are commonly reported in paranormal lore. Sources differ on what they're meant to be, with explanations ranging from ghosts to demons to aliens to time travelers to everything in between. They're described as humanoid shadows, sometimes with glowing eyes. Whatever they are, they're often considered to be a malevolent force.

Mary had some stories of her own. One day while she was upstairs, she heard a sound coming from the playroom in the basement. She described it as "the sound that the toy cash register makes." Kelly did indeed have a toy cash register in the playroom and upon investigation Mary discovered it was indeed the source of the sound. However, it was solar powered and the lights had been off when she heard the sound. No one else was in the room.

For understandable reasons, all three members of the family were a bit creeped out by the things they'd been experiencing and reached out for help. First, Mary reached out to her church. Their recommendation was that she should go to each room in the home where something had happened, pray, and tell whatever was there to leave her and the house alone.

She took their advice and began wandering the house to pray and attempt what fundamentally amounts to a lay exorcism. Things proceeded about as expected for a while, but when she reached the basement playroom, she heard a "loud growl," and then the toy cash register said "Thank you for shopping. Bye!" While that phrase was one of the toy's programmed responses, the entirety of the situation terrified Mary, quite understandably.

Knowing something had to be done, Mary next contacted a local psychic (whose name we'll not reveal). The psychic examined the house and explained that the home contained several spirits: two children, a man who'd choked to death after suffering lung cancer, a lady who owned the sewing machine cover the family was using, and a mentally disturbed woman (this was the one blamed for growling at Mary in the basement), as well as "several more spirits."

We need to offer a word of clarity regarding spirits identified by psychics. Psychic ability, quite simply, has not yet been scientifically demonstrated to exist. We're not going to argue here that there's necessarily no such thing, but if one relies on a psychic to identify a spirit, one is relying on an unknown practice to validate another unknown phenomenon. Therefore, while we do not immediately dismiss psychics' identifications as false, neither do we accept them as valid. Even if psychic abilities do exist, we maintain that not all who profess to be psychic actually have such abilities, and we've encountered several cases in which psychics have claimed to identify spirits only for our investigation to later debunk their claims.

On her own authority, the psychic reached out to TAPS for further assistance, who in turn contacted a local team (not Rocky Mountain Paranormal—not yet) to visit the home and investigate. This team stated they collected "several Class 5 EVPs" and were able to obtain more recordings in this home than in any other location they'd visited.

Again we pause for a moment of commentary. EVP refers to Electronic Voice Phenomena. Much paranormal lore suggests that spirits may use technologies to communicate with the world of the living. Historically, such items as spirits boards (Ouija boards) have been used. Modern times, though, require modern solutions, so many people capture audio recordings and then carefully listen to them for evidence that the spirits are speaking into the audio recorder in a manner that can't be heard without the recording device. One method involves recording white noise (static) and then listening for words. Another method involves rapidly scanning through radio stations and then listening for meaning in the snippets of words and sounds recorded.

In general, we find such evidence unconvincing because the recordings are made without adequate control for numerous psychological or environmental factors that could easily bias the results. We have developed protocols for doing this kind of research (though we tend not to on our investigations most of the time). Full details are beyond the scope of this book, but the gist of our protocol is that interpretation of EVP evidence must be conducted by an independent investigator who is held blind to the conditions surrounding the recording. These protocols were not followed by the group we're talking about.

With regard to the "Class 5" designation, there's no real agreement among paranormal investigation groups regarding classification schemes for much of anything, EVP included. What we believe this group was referring to, however, is a classification scheme in which the quality of EVP recording is rated on a scale from 1 (poor quality; perhaps unintelligible) to 5 (superlative).

Discussing the case on the phone with Judith, this ghost hunting group claimed that they'd "released" the spirit of the mentally disturbed woman who'd growled at Mary in the playroom as well as the man who'd choked to death. Further, they explained, the home was a "portal" (presumably a portal to the spirit realm, though they didn't elaborate much) and that the "energy" could not be stopped but that the malevolent entities had been removed.

By this time, the family still felt they needed more help, so they eventually reached out to us (on an additional referral from TAPS). Meanwhile, the original group returned to the home to attempt a "sealing of the portal." Even while the family was in communication with Rocky Mountain Paranormal, the other group spent a Saturday at the home, collecting audio and video recordings for about eight hours. Every member of the team, they said, experienced something. "As far as the Portal goes," however, they said "we could not seal it completely and believe that there are roads leading into the main portal so there is more work to do....There are many souls who will remain here for one reason or another. A lot of children I am told by the EVP."

A follow up message from the group said, "There is still a lot of activity. Not so much whispering but other noises. The Portal has not sealed because there are roads leading into it....We are working on about four and a half hours of evidence....I had been sensing two new spirits, young men, or teenagers that needed to get a message through. So I asked our sensitives to come back to see what they could sense."

For the record, "sensitive" is essentially synonymous with "psychic."

These alleged sensitives returned to the house and issued another report of their own: "We went into the play room and sat on the floor. Almost immediately, we felt the presence of a young man named Joseph and another naked Erik. Both had died in alcohol related accidents. When we asked them to show a sign of their presence, the DVD play [sic] in the playroom immediately started to buzz almost in Morse code. We would ask a question and they would respond using the DVD player. There were no cell phones, the lights were out, and the DVD player and TV was turned off. It was quite amazing. When we left the playroom and went to the computer room, we turned the TV on to a snow channel to see if we could get more responses. As we were talking, a white flash came from the playroom and when we went in, we found they had turned the TV off. Therefore, there is a lot of activity in the house. No portal has been closed a when out [sic] medium is present there is a lot of contact. We stirred up things a bit. Do you have a KII meter?"

There's a fair amount to unpack there before we continue with the story. First of all, for the record, all of these communications are presented verbatim with grammatical and spelling errors intact. It's not that we're trying to embarrass anyone, but we want to present a complete and accurate record of the case, down to the smallest detail. Second, though the psychics reported that the DVD player sounded like it was communicating in Morse Code, they made no attempt (to our knowledge) to determine if it actually was Morse Code or to decipher the message. Finally, with regard to the last line, the K-II (letter K and Roman numeral for the number 2) is a commercially available EMF detector. It's commonly used by ghost hunters, but is not, in our opinion, the most accurate or reliable EMF meter on the market. Indeed, even in its (admittedly low) price range, there are several other models we prefer.

We were in communication with the family during all of this, but it was at this point that the family decided they needed to bring us in as soon as possible for additional on-site investigation. We scheduled an investigation and received the following confirmation from Mary: "That sounds good to me. My husband is patient and he will be going to the neighbors to drink scotch and relax. The last team got many EVPs and about 42 instances of doors closing, footsteps, knocks, and assorted bumps in the night. Their sensitive got a whole lot of interaction by my ghosts wanted me here so your idea of having me here is a good one. You will get an enormous amount of EVPs so be prepared. You can bring as many sensitives as you want. My daughter and I will be able to contact and feed off of your sensitive. Seems to work quite well. We are not trained enough to get all of the messages on our won. I will try to tell what you are feeling when you are here. I usually can pick up when they are here and am good with names but I need help. I am anxious to come experience my home; it will amaze you. See you [date redacted]."

On the day of the investigation, we arrived at the property at around 6:00 p.m. and met with the residents to check in and see if there had been any recent activity. Both Mary and Judith were present as well as (unexpectedly) a neighbor and her daughter.[1] They didn't have anything new to report, so we began setting up.

Given the amount of activity which took place throughout the house, our setup was extensive (though focused primarily on the basement playroom as it seemed to be the nexus of most of the activity).

We placed three video recorders attached to remote monitors (this was before we used surveillance cameras and a DVR system) in the dining room facing the first floor living room, at the top of the second floor stairs looking into the first floor living room, and at the end of the basement hallway looking toward the stairs (this last camera was moved halfway through the evening to the middle of the hall). We also had two video cameras connected to a computer for remote monitoring. These were placed in the master bedroom and the playroom.

Two remote thermometers were placed in the master bedroom and the first floor bedroom.

For EMF detection, a cell sensor EMF meter was placed in the basement playroom. Additionally, we regularly swept the entire house with both AC and DC (natural) EMF meters. We did not use the aforementioned K-II model (though we have some of them in our tool kits) but instead used the more-reliable Tri-Field models.

Microphone and audio recorders were placed in the playroom, master bedroom, and living room.

Finally, we used high quality still and video cameras throughout the evening to document the investigation.

During the setup process we took baseline readings on the EMF meters and thermometers. Readings on the EMF meters were particularly high (15-25 milligauss) in the backyard. We determined this was due to the underground power and cable TV lines. Throughout the home, EMF readings were consistently high throughout the evening, but did not fluctuate much as the evening wore on. The first and second floors had unshielded wiring and the walls of the covered patio contained the main power bus and breakers for the house. All of these contributed to the EMF readings. Readings in the master bedroom never exceeded 1 milligauss.

In the playroom specifically, we noticed that we obtained high (10+ milligauss) EMF readings when using the meters above a height of about five feet off the floor. We determined this was due to the wiring and lighting in the room. Below five feet, readings were substantially lower. This is an important point. Electromagnetic fields are often quite strong near a source, but levels drop off significantly with distance. Specifically, electromagnetic radiation follows an inverse square law: the strength of a field is

---

1   As a rule, we try to discourage people from inviting guests to our investigations. Too many people can alter the conditions in the house and risk contaminating data.

inversely proportional to the square of the distance from the source. In practical terms, this means you generally have to be relatively close to an EMF source to measure it.

Temperature readings were mostly consistent throughout the evening at an average of 72 degrees Fahrenheit. Temperatures did drop over the course of the evening in the covered patio, but this was due to the falling evening temperatures and the lack of insulation in that room.

Throughout the evening, Mary and her neighbor insisted on going to the playroom and trying to communicate with the ghosts. This is not our usual protocol, but we didn't try to stop them, as long as they gave us at least some times during which we could monitor the house in silence. When they went to the playroom, they used the K-II meter that had been suggested by the previous group and claimed they were not only getting rather different readings than we were but that the ghosts were using the device to answer their questions.

Because these devices can be sensitive to how they're handled, we suggested they should put the meter on the floor when they asked their questions. When they did this, there was no response. This led us to think the responses may have been due to their handling of the device rather than a paranormal source. On further investigation, we discovered that when they were asking their questions, they were moving around the room. When they would move, the device would light up.[2] The device they were using during this experiment had a power button that required constant pressure in order for the meter to stay on. Any slip of a finger or inconsistent pressure could easily account for the inconsistent readings they obtained.

At the end of the evening, we hadn't experienced anything unusual in person, but we had a lot of video and audio to review. As we were packing up, the family provided us with a CD containing the EVP recordings from the previous team, and we added that to our material to be reviewed. To say the least, we weren't impressed by its contents. Though it was claimed to contain a high number of EVP recordings, we were unable to determine whether or not anything anomalous had actually been recorded because the ghost hunters conducting the EVP session never stopped talking. A ghost could have walked directly up to the microphone and recited the Gettysburg Address into it and we'd never be able to know. On the occasions we've been able to obtain copies of other teams' EVP recordings over the years, this has been a consistent problem, so we'd like to take the opportunity to advise people, if they're going to try to talk to ghosts with audio recorders, to at least give the poor ghosts a chance to get a word in!

On our own recordings, we didn't find anything unusual.

That left us in a bit of an awkward position because we had to contradict what so many other people had said. True, we can point to plenty of flaws in their methodology,

---

2   K-II meters feature a series of lights rather than a screen or dial, and so don't provide specific values for the strength of the electromagnetic field.

but we still hoped we'd be able to experience some of the claims in question.

We do have a few ideas worth mentioning. We did notice high levels of electromagnetic radiation in the playroom and some other parts of the house, including all of the locations where the homeowners had reported paranormal activity. While hallucinations or paranormal experiences directly caused by electromagnetic fields are relatively rare and typically occur only in those with particular neurological predispositions, we can't entirely rule EMF out as a cause of at least some of the experiences.

Once paranormal lore had begun to develop in the house, the confirmations provided by other ghost hunting groups may have further developed the idea for a group of people already predisposed to believe in the paranormal claims.

However, while electromagnetic fields and psychological biases are a possible explanation for the claims in this house, we're far from convinced. We would have liked to investigate a lot further, but unfortunately the clients eventually dropped out of communication with us, so we haven't had the opportunity to revisit the house nor do we know whether paranormal experiences persisted.

20
The Most Haunted House Ever

Continuing our discussion of cases in which we've responded to a call after other groups have already investigated the location, we present a case that took place in a house described by the clients as one of the most haunted houses ever, with implications of possible demonic activity.

As you'll see, our findings were again a little bit different than those of the prior teams to investigate the property, and further illustrate why we emphasize the importance of contacting professional paranormal claims investigators instead of fly by night ghost hunting operations.

Figure 20.1. Artist's rendition of the most haunted house ever. Artwork: Aaron Bordner.

This will be a fairly short story, and one of the shorter chapters in this book, because the information we were told in advance of our investigation was limited, and once we arrived for the on-site investigation, we made a discovery that necessarily cut our work short.

The home was occupied by a couple who we'll call Bernard and Susanne as well as their young children. The case began when they started experiencing a wide variety of paranormal phenomena. Odd sounds were heard coming from the kitchen. Their home surveillance system picked up voices even when no one was present. Photos captured by the same surveillance system depicted what the residents believed to be ghosts. Things got so frightening they eventually began to suspect perhaps a demon, rather than mere ghosts, was causing the increasingly violent disturbances.

Multiple other ghost hunting teams were called to the location before Rocky Mountain Paranormal got involved. These teams recorded what they described as "threatening" EVP messages. Ghost hunters' own photographs also depicted ghostly images. Most disturbing of all, some of the other ghost hunters reported that Bernard had begun to exhibit violent tendencies, which they took as confirmation of demonic activity in the home.

It was at that point Rocky Mountain Paranormal got involved. We began by examining the EVP recordings and ghost photographs. Universally, these were easily explainable by mundane causes. Some of them didn't even seem to depict anything at all, and those that did were the product of known optical or auditory phenomena.

The sounds they'd heard coming from the kitchen were a bit more difficult to explain, but we eventually found the culprit: a combination of the home air conditioning unit and the refrigerator caused the floor to vibrate.

While we were there, though, we became increasingly disturbed by Bernard's behavior. Other teams had already noted that he had begun to exhibit increasingly violent behavioral tendencies, which they attributed to a demonic presence. We noted the aftereffects of one of these behaviors ourselves. During our investigation, Bernard was wearing a cast on his hand, and we were told he'd recently punched a hole in a wall in the house in a fit of temper.

Our interview suggested a different cause than demonic activity. Bernard told us he'd been experiencing abnormal levels of stress both at work and at home. While we're not clinical psychologists (and even if we were, we would not be acting in that capacity during an investigation), we suspected that these stressors might be the cause not only of Bernard's behaviors but of many of the paranormal claims.

Stress, trauma, and paranormal experience often coincide with one another. We're always careful to recognize that this is true whether one believes in the paranormal or not. Lore suggests that stress can invite malevolent paranormal entities into one's home. Psychology also suggests that stress can cause one to hallucinate or believe in paranormal entities. Either way, the cure is the same: put your life in order, and the paranormal experiences tend to stop.

Given the tenuous psychological situation of the family, we chose to stop our on-site investigation early pending psychological review of the case. We sat down with Bernard and Susanne and gave them several recommendations. First, we suggested they should both seek professional counseling for stress and anger management to get their psychological state in order. Meanwhile, they should search for normal explanations for their experiences before assuming a paranormal cause.

Most importantly, they should immediately stop working with ghost hunters who were making the situation worse by excusing violent behavior as the result of demonic activity.

That's not to say one can never or should never call paranormal investigators. Calling us was actually a good idea. And we're not the only ones who do good work. But the particular groups this family had chosen first were looking *only* for paranormal explanations, failed to employ basic critical thinking techniques, and ended up exacerbating the situation. If indeed it was stress, rather than a demon, causing Bernard's violent outbursts, then telling him and his wife that there was a demon in the house surely increased the stress, potentially causing a positive feedback loop that could have ended catastrophically. Fortunately, we were able to intervene in time.

In this case, the clients were responsive to our suggestions. They thanked us for our time and advice, agreed they would at least consider natural explanations in addition to (and preferably before) paranormal ones, and most importantly, they agreed to seek psychological counseling to help with the anger issues.

We asked them to contact us if the paranormal phenomena continued or they needed any further help. As far as we know, the counseling did the trick and they stopped experiencing anything ghostly or demonic.

The most haunted house ever turned out not to be so haunted after all, at least in the traditional sense. An argument could be made that the human brain is the most haunted place of them all, and this case is a perfect example of that.

21
The Demon Made Me
Beat Her Up

As a follow-up to the last chapter, we continue our exploration of "the devil made me do it" as an excuse for violent behavior. It's far from the most common type of case we encounter, but it is common enough to merit our attention. This case, which took place at a home near Denver International Airport (itself home to plenty of paranormal stories, though those are a topic for another time), was referred to us by another paranormal investigation group who quickly realized they were in over their heads.

Figure 21.1. Artist's rendition of the demon made me beat her up. Artwork: Aaron Bordner.

"The devil made me do it." It's a phrase offered lightheartedly by children as a way of excusing minor indiscretions. It's also been used as a defense to criminal activity, both in horror stories and, from time to time, actual legal cases. The third volume in the popular *Conjuring* franchise of horror movies is even called *The Conjuring: The Devil Made Me Do It* and fictionalizes a supposedly true story in which self-proclaimed demonologists Ed and Lorraine Warren consulted on a case in which an accused murderer attempted to argue demonic possession as part of his legal defense.

Unfortunately, we occasionally happen across such cases in our own paranormal investigation work, and they often present substantial ethical challenges for us. We'll begin by discussing this particular case, which will be brief because the investigation didn't last long, and then offer some more general thoughts on how these cases should be approached.

The family involved in this case consisted of a mother and father, both in their 30s, who we'll call Ed and Lorraine, and a 15-year-old daughter we'll call Chelsea. They first contacted a different paranormal group in town (do you see a pattern in these cases yet?) with a variety of claims they believed were demonic in origin.

Ed claimed he'd heard people (not his family) talking throughout the house, and especially in the home's raised kitchen which was attached to the living room. All members of the family reported sounds of movement in the basement. Their dog even refused to go down there. For his part, Ed felt he was occasionally taken over by a demonic force, which had all of them completely terrified.

The group they'd first contacted conducted an EVP session and said they were so terrified of the results—including sounds of growling and demonic voices threatening both the family and the paranormal crew—they immediately knew they were in over their heads and referred the case to us. Based on their description, we found the case concerning enough to immediately schedule a home visit (eschewing the usual period of preliminary communications).

During our tour of the home, we noticed several disturbing things. First, there were several holes in the walls that hadn't yet been patched. Worse, Lorraine's arm was in a sling. We inquired about these matters and both Ed and Lorraine confirmed our worst fears: Ed claimed he was sometimes taken over by a demonic force, during which episodes he would become violent. He admitted both to punching holes in the walls and to pushing his wife, causing her injured arm.

At this point, we immediately stopped the investigation. We told them that we'd be more than happy to continue but that before we could continue to participate, we'd need their assurance that they were pursuing professional help from a psychologist or therapist. We told them we weren't necessarily denying a paranormal aspect to the case (and we weren't) but we needed to make sure someone was helping with the family dynamics to get them through the situation before we could go any further.

Ed and Lorraine were both outraged by our request and demanded we leave their house, which we did. We tried to continue communications with them, but they refused to reply to any of our letters or phone calls.

Obviously, this is not the outcome we wanted, for multiple reasons. For one thing, we genuinely do want to investigate the paranormal claims whenever someone approaches us. More importantly, we wanted to make sure they got the psychological help they so obviously needed, whether or not there was actually a demon causing the disturbances. Because they broke off communication, we don't know the final outcome of the case or whether they eventually managed to seek help or work through their issues.

Since this case, we've developed protocols for dealing with these kinds of cases that present remarkable ethical challenges. For instance, we find ourselves in an awkward position when caught between ethical requirements to keep clients' interactions with us confidential and to report suspected violence or abuse. We've given these matters a lot of thought and consulted with ethical guidelines in other professions and developed methods for dealing with these situations.

First, there's an important note of philosophy regarding such cases which we share with the Catholic Church. The Church absolutely does believe in the possibility of demonic possession and will perform exorcisms. In fact, you'd be surprised how common such rituals actually are these days. But they will not do so without obtaining permission from the local bishop, and that permission will not be granted without a psychological evaluation. They do this for two reasons. They want to ensure that they're dealing with a genuine case of demonic possession rather than someone who is mentally ill and should be treated by a psychiatrist, of course. But also, they realize that even if they do believe it to be a genuine case of demonic influence, there are benefits to the client in having a psychological or psychiatric professional help process the terrifying and confusing and stressful situation to ensure that no additional psychological harm is done.

Guided by that philosophy, our current protocol for dealing with "devil made me do it" cases is similar to the protocol followed by the Church. If we believe children are being abused or anyone is in imminent danger, we will immediately report the case to the proper authorities to prevent tragedy, following guidelines established by law.

For cases of suspected child abuse or neglect, the State of Colorado, in C.R.S. 19-3-304, identifies numerous professions considered "mandatory reporters" who can be penalized by law for failure to notify the relevant authorities when they discover or form suspicions of abuse or neglect of a minor. Paranormal investigators are not specifically listed in the law, but we consider ourselves closely enough aligned with some of those professions that we take guidance from the law even though we may not be strictly bound by it.

The law provides for similar mandatory reporting of abuse of elderly, at-risk, or mentally impaired adults for a similar list of professions. Our protocol likewise takes its guidance from this law even though paranormal investigators are not explicitly mentioned in the law's text.

For adults not considered at risk, reporting is not mandatory. In those cases, we will protect client confidentiality unless we suspect a life is in imminent danger, though we are more than happy to help any victim escape a dangerous situation or find professional care or resources.

If the case hasn't yet gotten to the level at which reporting to the authorities is necessary, we will not immediately contact law enforcement, but instead insist upon bringing in outside help from clergy. Not necessarily Catholic clergy unless the clients themselves are Catholic (though we did take some inspiration from how the Catholics deal with these cases), but rather clergy from the clients' own religious tradition. Importantly, however, we insist upon working with clergy who share that insistence upon professional psychological oversight of the process.

We will continue to investigate such claims, as long as it is safe to do so, working in concert with the clergy we've brought in. It's our hope that by working with clergy first and psychological professionals second, we can avoid the kneejerk reaction (as we received in this case) of people thinking we're calling them crazy or accusing them of something. That's not the case. Our only interest in these cases is in making sure the clients get the help they need, whether it comes from a doctor or a priest or whatever else.

Even if one does not believe in the reality of demonic possession or rituals of exorcism, there's an additional reason we want to work with clergy in these cases. If the client *does* genuinely believe that his or her behavior is the result of demonic influence, then whether it's true or not, the ritual can be successful in ending the violent behavior, even if for purely psychological reasons. In fact, we suspect (though to our knowledge it has not been formally studied yet) that exorcisms may prove more effective than other forms of psychological treatment in such cases.

We remember this case as one of our greatest failures, but even in failure we have learned important lessons that continue to guide our current work. It's our sincere hope that the family involved eventually got the help the needed and worked through their issues.

22
The House by the Cemetery

It's not our habit to become directly physically involved in the paranormal activities in our cases, but it happened involuntarily in this case as it involves our founding member Bryan getting slapped by...something.

The case in question took place at a house by a cemetery near Fort Collins, CO. It was a multi-generational home and several members of the family were, by the time they contacted us, growing increasingly terrified of their own home.

Figure 22.1. Artist's rendition of Bryan getting slapped by a ghost. Artwork: Aaron Bordner.

The story here begins with a rather unusual family situation. The house was a multi-generational home consisting of six people (as usual, with their names changed to protect their privacy):

- Michael—the father,
- Cindy—the mother,
- Stephanie—the eldest daughter, aged 17 years,
- Leah—the younger daughter, aged 8 years,
- Kaley—the granddaughter (Stephanie's daughter), aged 5 years, and
- William—Stephanie's boyfriend (Kaley's father), age unknown.

Take a moment to work out the arithmetic, and you'll understand why we mention the family situation was a bit on the unusual side. Whether that has any bearing on the paranormal experiences the family claimed remains an open question. However, to their credit, they did seem like a loving family.

Their home was built in the 1970s and, as the title of this case might suggest, neighbored a cemetery. However, as far as we were able to determine in our background research, none of the previous occupants ever reported paranormal activity. Either the paranormal claims began with this family, prior occupants lived with the phenomena without ever telling anyone about it, or we simply weren't able to find the reports. Any of these is possible, though we emphasize that we're fairly thorough in our searches. It would be nice if there were some kind of national or international searchable database of paranormal claims, but neither we nor anyone else has yet figured out how to build such a thing while still accounting for client privacy and preventing being overrun by fraudulent entries.

When they contacted us, the family reported numerous claims. Unexplained sounds came from all over the house. Young Kaley once had the blanket "fly" off of her bed. Multiple family members said they'd seen unknown persons in the home late at night, and three of them witnessed a book moving across the room by itself.

Around three months before they called us in to study the house, the bulk of the family had been vacationing in Florida, leaving William and Stephanie alone at the residence. One evening during this period, at around 11:00 p.m., the young couple heard footsteps and knocking outside the home. Upon their investigation, they found no one present and nothing out of place. Identical events occurred every evening for several days. Eventually, as time went on, the noises started to sound like they were coming from inside the house.

After a while, the rest of the family returned home. That first night back, Cindy heard what she described as a "loud knocking and pounding" sound on the wall of her bedroom, which was closest to the garage. When she got up, she discovered the rest of the family had similarly awakened and were all trying to figure out where the sound had come from. None of them could identify the source. This persisted for the next several weeks, with the sounds increasing in both volume and quantity. Again, they seemed to be moving in from the outside and occurred more regularly inside as time went on.

Some weeks later, things again intensified. The family began seeing "shadows in the middle of the room and red lights floating through the room." Certain rooms began to feel cold.

They told us the vast majority of the paranormal activity was concentrated in a single bedroom in the house. So where has the *entire family* chosen to sleep at night? We'll give you three guesses, but you won't need them.

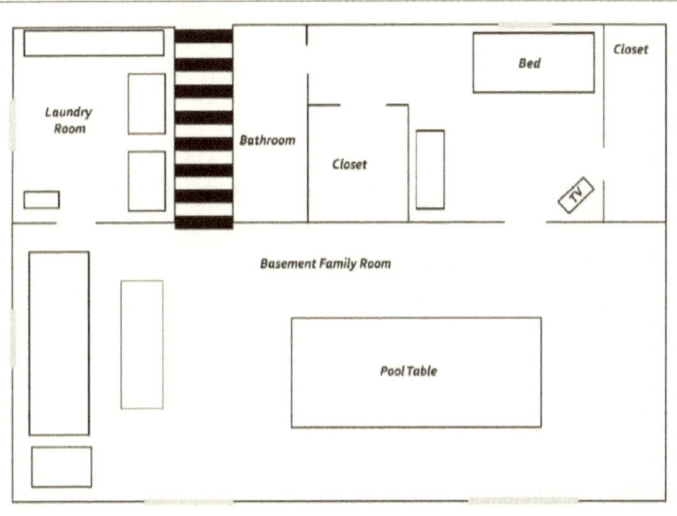

Figure 22.2. Basement floor plan (including haunted bedroom) of the house by the cemetery. Image: Bryan Bonner.

One night, as they all slept in the same room, again at around 11:00 p.m., the noises resumed. This time, though, they heard what sounded like someone or something walking down the hallway. Stephanie, attempting a sort of spiritual defense, removed a crystal she kept on a necklace and placed it on a book near the room's door. Several minutes later, the book and crystal moved across to the other side of the room, witnessed by the entire family. The noises persisted and grew louder. Bedsheets started to pull down by themselves. Stephanie's nightgown started to pull itself up. Noticing this, William grabbed it to pull down the cloth, and later said he felt a "pressure pushing [his] arm away from her" and felt his own blanket tighten around him until he moved away.

Later that night, the family heard a handheld video game system that was in the room power itself up. A popular game started playing on the device. No one was in the room at the time. Stephanie went to investigate. According to her story, when she picked up the game and looked at the display, she saw the reflection of two people on the screen even though she was the only one in the room.

As you can plainly see, this case was a doozy. All of us are big-time horror fans and some of us even write in the genre. Even we would struggle to come up with a better ghost story than the one related to us by this family.

After that night of terrors, the family decided, quite understandably, to seek help, and they reached out to Rocky Mountain Paranormal. We received the following email from Cindy the next day (edited for brevity): "Hi! My name is [Cindy] and I live in Ft. Collins, CO. If you could please direct me to someone who could help me with a 'not so nice something' in my house. As of last night, this 'thing' has become more forward and doing things that are frightening to my children…. Something is down in my basement, and I cannot figure out what it is and don't know where to turn for help. Please, if you can direct me to someone who is close to Ft. Collins, CO and pleas [sic] know, I don't have any money, but I am in dire need of assistance. My husband thinks we are all nuts, so is of no assistance at all. It would be GREATLY appreciated."

Of course we were more than happy to help. Fort Collins may be a bit of a drive for most of us, but we're not afraid of a short road trip when a case presents itself. We wrote back to explain that we'd do whatever we could and that we never charge for our services (indeed, we consider charging for anything beyond basic expenses on a paranormal investigation case to be a red flag that you might want to consider a different group to help with your problems).

Before the investigation, because the family was collectively frightened of their own home, we arranged an interview at a local restaurant, where they told us the basics of the case in more detail. One of them, Stephanie, initially seemed somewhat resentful of our investigation. When we tried to resolve this tension, some hostility between family members surfaced, and we led the conversation toward a discussion of family dynamics. Things were somewhat volatile within the house and included an impending divorce, drug abuse, and teenage pregnancy. Digging into family secrets gives us no pleasure, but all of this information becomes quite relevant in paranormal investigation.

Stephanie admitted to practicing witchcraft (her own word, which we determined in this case to refer to a Wiccan belief system) and said she felt responsible for inviting something into the home, which at least partly explained her hesitation regarding the investigation. Once we'd gotten the information (and all the details of the story as told above), we scheduled an on-site investigation.

We arrived shortly before 9:00 p.m., took a tour of the house and started setting up our equipment and collecting baseline data. Around 10:00 p.m., we had to remind the family that they needed to stick to their normal routine as closely as possible and asked whether this wouldn't be about the time they'd be going to bed. All of them seemed reluctant to do so, though, and no one made any actual moves in that direction. They finally decided that they'd only be willing to go to sleep if they did so in shifts, and with our team monitoring. We're not quite sure what they thought we'd be able to do to protect them if the boogeyman did decide to show up in the middle of the night, but we did not share these thoughts and agreed to help them keep watch.

At 10:45 p.m., there was a fresh bit of excitement when one of the motion sensors we'd placed in the bedroom (the same bedroom where the worst of the activity had taken place) went off. We rushed into the room to see what was going on, but this turned out to be a false alarm. All we found was a kitten, rather terrified by the sound of the alarm, who had wandered into the room and set off the motion sensor.

At 11:50, Leah went into the bedroom and tried to sleep. After a few minutes, she jumped out of bed and said, "that scratching is starting again." We successfully recorded a scratching sound coming from a concrete foundation wall on the west side of the room. Concrete foundation walls tend not to make those noises under ordinary circumstances, and we searched long and hard for an explanation but could find none. When we played back the recording of the sound, the family confirmed it was the same noise (or at least one of the same noises) they'd been hearing. Mice or other critters are a common go-to hypothesis for scratching sounds, but because this was a concrete foundation, we ruled that out.

At 12:13, the motion sensor in the bedroom closet went off. This time, there was no cat. No one had been moving around. To this day, we don't know what set it off.

At 12:15, Stephanie and William took a shift in the bedroom. Thirty-five minutes later, Stephanie came out of the room and explained that she just couldn't take the "feeling" in there. Leah took a shift in the room, which was uneventful for the next hour. At the end of her shift, Stephanie and William went back in to try another round. Just thirteen minutes in, Stephanie said she could see something in the room. The EMF meter we'd placed in the bedroom made an audible noise (it was a model that emits a sound when it detects an electromagnetic field above a certain threshold) for about 20 seconds. Both Stephanie and William exited the room, claiming to have trouble breathing in there.

Once they'd calmed back down, Stephanie and William said they'd go back, but only if RMP member Bryan went in there with them. Still unsure what Bryan would be able to do in defense against evil forces, we agreed with the modified protocol. Bryan sat on the far end of the room as the young couple went to bed.

Several minutes later, Stephanie said she felt like there was something at the side of the bed. At the same time, Bryan felt like he got slapped by a ghost or some other unseen force or entity. He described the sensation as a "light slapping" on his face. The slap was followed by a noise from the wall about two feet or so above Bryan's right shoulder. While all of this was occurring, the EMF meter again started to emit its chirp. All of this lasted for about a minute.

Figure 22.3. Rocky Mountain Paranormal investigator Bryan in bedroom with sleeping family members (cropped out). The arrow indicates the approximate location of the sound Bryan heard. Photo: RMP archives.

The room was in total darkness, so none of the people inside could see anything. However, the rest of the team outside the room were monitoring the room using a camera with infrared night vision. Bryan said "I hope you guys can see something. It feels like something is slapping me." There wasn't anything on the video.

At the end of this episode, Leah and Cindy took a shift in the room. Twenty-five minutes in, Cindy said she was cold and that it felt like something was in the room with her. She left a few minutes later. Stephanie and William took back over in the room, with another of our team members joining them. Stephanie said she could see what appeared to be a silhouette of a man wearing a cape near the closet door. Our team member also confirmed that there seemed to be something in the room, though it was so dark it was impossible to tell who or what it might be. The video didn't reveal anything.

We concluded the investigation at about 5:00 a.m. It had been quite an evening. Alas, we weren't able to answer all of the family's questions. Many of the things that occurred during the night can be dismissed as merely the product of all of our imaginations playing tricks on us. When family members had "bad feelings" about the room, that could easily be dismissed as mere anxiety. Even Bryan getting slapped could be the result of psychological rather than paranormal phenomena. After spending an entire evening listening to ghost stories and then sitting in the dark in the supposedly haunted room, it's possible for one's brain to play some dirty tricks. But the fact that we can't conclusively rule out either the psychological or the paranormal explanation makes this one of our intriguing stories. Ultimately, the only thing we can rule out conclusively is the possibility that something physical slapped Bryan's face because if it had, it should have been seen on our camera.

On the other hand, some of the other happenings like the EMF meters and motion detectors going off or our recording of the scratching sound can't be so easily dismissed and we weren't able to find explanations.

Members of the family requested a blessing. As we're not clergy, we were unable to provide it but we reached out to some people in our network and arranged for a blessing to be performed and followed up to ensure it was done. Because the family dynamics we observed were rather volatile we recommended psychological counseling to work on their issues and to help them deal with the stressful paranormal situation. They agreed to do so. We reached out a few weeks later and they said there'd been no further activity.

So where does that leave us? Aside from a developing hypothesis that ghosts enjoy slapping Bryan for some reason, we really don't have much to say. If everything had been psychological, maybe the counseling helped and made it all go away. But that still leaves the question of some of the physical phenomena we recorded. If it was paranormal, maybe the blessing worked, but we always try to exhaust every possible natural explanation before assuming the paranormal one. We really don't know what was happening in that house.

Getting comfortable with "I don't know" as an answer at the end of an investigation is a very important requirement for any paranormal investigator.

23
The Toilet Ghost

This is one of our favorite stories. Despite the humorous name (which we use with the blessing of the clients involved who were quite pleased when we assigned the title), it provides a great example of a private investigation in which everything went right. In fact, it provides such a textbook example of a private investigation that we often use it (or elements from it) in our educational lectures.

Figure 23.1. Artist's rendition of the toilet ghost. Artwork: Aaron Bordner.

Sometimes people call us because they have experienced just one or two weird things and they want our help figuring them out. Other times, they have so many experiences that there's a temptation to think even if some of them are explainable as mistaken natural phenomena, at least some of them almost have to be paranormal in origin. This case was one of the latter. The family involved had been experiencing a large and diverse collection of phenomena and were growing increasingly fearful of their own home. As always, we were more than happy to assist.

Our clients had moved into their home about a year before they contacted us. It was a single-story ranch style home with some additional commercial buildings on the property in an extremely rural area, bounded by two creeks. Residents included a father who we'll call Wayne, a mother, Charlotte, and two daughters, Jessica and Melissa. They also kept a small dog and a bird as pets.

When they first contacted us they said the entire family had experienced strange events. Often, this is not the case—you may have noticed in some of our previous case files that it's common for some family members to believe in the paranormal and others to resent even the presence of an investigation team. In this case, though, the family were unified in their concern.

Their initial reports when they first contacted us said that their youngest daughter, Melissa, was having trouble sleeping and seemed afraid of the dark. Not an uncommon thing for young children, though in this case it rose to atypical levels. Melissa insisted on sleeping with the lights or television on, reported seeing a figure in her peripheral vision which would sometimes sit on the edge of her bed, and had recently taken to insisting on sleeping in her sister Jessica's bedroom at night.

Additionally, Charlotte described sometimes hearing sounds akin to a radio or television in another room when she was alone in the house, but finding no devices turned on whenever she would investigate. Sometimes the sounds resembled a conversation between two or more individuals, though she couldn't make out what they were saying.

Wayne and Melissa both reported hearing footsteps in the hallway and the sound of door handles rattling. Charlotte added that she'd seen the bedroom door moving.

Finally, they said a ghost was flushing one of their toilets at odd intervals, particularly during the night. We immediately suggested an experiment they could perform even before we visited the site, but we'll refrain from discussing it until we get to the discussion of our investigation a little bit later.

Based on their communications, we assumed there were even more stories than what they initially told us so scheduled an on-site interview to begin the investigation process. Meanwhile, we advised them all to keep diaries of everything they experienced so we could discuss them during the interview. They did so, and they did it correctly: they each kept a diary but they didn't share their diaries with each other until we interviewed them, so as to avoid contaminating or influencing each other's stories.

During our interview, we learned more about the family and their experiences.

They'd lived at the home for about a year, though they also claimed to have some paranormal experiences at a prior residence. Those never rose to the level of what they were currently experiencing, but they did wonder as to whether something was following them personally rather than specifically haunting the physical location of their residence.

Diaries and our interviews with each family member confirmed many of the claims they'd made in their initial letters and also added some additional ones and clarified a few details.

Wayne reported hearing footsteps and the sound of the door handle to his bedroom moving. Charlotte reported hearing doors opening, footsteps, and the aforementioned snippets of indecipherable conversation. One day when she came home from work late, she saw what looked like a person standing by the back door to the garage. When she turned on the lights, no one was there.

Jessica, the oldest daughter, described hearing door handles moving. Melissa, the youngest daughter, seemed the most terrified of the family. She reported hearing footsteps, door handles moving, and seeing a shadow figure in a hallway leading to the kitchen as well as multiple encounters with a shadow figure (perhaps the same one) in her bedroom. Several times, she called her mother at work because she was afraid of something she'd experienced. She believed she was being harassed by the devil himself and that the ghosts had wanted to kill her under the stairs of her old house.

Over the past year, the home had been vandalized on some occasions. Knobs on the outdoor water faucet had been removed, the exterior lights had been broken, someone shot the mailbox (apparently at close range with a .45 caliber round), and signs on the property had been cut in such a way that they would fall forward. We chose not to focus on the elements of vandalism in our investigation because we felt those were unlikely to be related to the paranormal claims and were a matter best left for law enforcement to handle.

The experiences reported by the various family members were not identical, but they did share many common elements. Shared delusions or hallucinations are not unheard of in this line of work, but the similarity between the stories lends some degree of credibility to the claims and points (not absolutely, but to a certain extent) toward an external cause rather than a psychological one, for at least some of the phenomena.

Ultimately, we divided up the experiences into a list of primary claims:
- The toilet flushing by itself,
- Footsteps walking down the hallway,
- Door handles moving,
- A lampshade moving with nobody near it,
- Feelings of being followed outside the home at night,
- The youngest daughter being afraid of ghosts/entities in her room,
- Seeing apparitions, and
- Hearing voices or conversation.

Toward the end of the interview, we decided to conduct what's known as a suggestibility test on the family members. We suspected that even if some of the phenomena turned out to be genuine, it was likely that at least some others might be the result of suggestibility and the family members' belief that their home was haunted. Psychologists have developed suggestibility tests as a part of hypnotherapy, of which some of our members are certified practitioners. Note: we were not trying to hypnotize anyone nor to use hypnotherapy as part of our protocol. Rather, we were interested only in using the one specific aspect of hypnotic theory to determine relative levels of suggestibility among the family members. Hypnosis itself is on the boundary of being a paranormal claim. It's poorly understood and psychologists are divided on the utility of hypnotherapy as a clinical practice. Almost universally, though, scholars agree on two things: hypnosis is a real phenomenon, and most of the more extreme claims made about it are exaggerated or false.

Use of hypnotherapy in paranormal research has sometimes involved what's called hypnotic regression therapy, used to unlock repressed memories. We do not endorse this practice, as it's far more likely to implant false memories than to unlock genuine ones. We recommend reading the works of Dr. Elizabeth Loftus for more information on this topic.

Perhaps unsurprisingly given her claims were the most developed and frightening, we determined that Melissa was the most suggestible of the family members, though all of them scored above average on our suggestibility scale. In light of this, we told them to be cognizant of the fact that at least some of their experiences could be the result merely of their belief that their house was haunted. Still, we wanted to come back for more investigation and scheduled a date to do so.

On the date of our investigation, we arrived around 8:00 p.m. Both daughters and a house sitter were present. Wayne and Charlotte were not at home at the time (we knew in advance this was going to be the case and had their blessing to proceed in their absence). A brief interview revealed that they'd been experiencing more of the same kinds of activity in the meantime. The only new claim was that Melissa said a picture frame mounted to the right of her bedroom door had swung by itself, but she admitted she couldn't be sure whether this was a new paranormal claim or whether she'd just accidentally bumped it without noticing.

While we set up our equipment, we asked everyone to go about their business as normally as they could and do their best to pretend we weren't even there (an impossibility when a crew of strange people are loading crates of scientific and surveillance equipment into one's home, but it's worth at least trying).

For our base of operations, we selected the covered patio area. No paranormal activity had been reported there, so we figured we could use that location to monitor remotely while staying well away from the family's normal activities. With that established, we began setting up our gear.

Thermometers were placed at the base of operations and in the hallway near the entrance to Wayne and Charlotte's bedroom. One seismometer was located in front of the door to the master bedroom. Additionally, we used another seismometer at multiple locations during our regular sweeps of the house. Two stand-alone microphones were placed in the house, in the family room and in the hallway near the master bedroom.

Most of our monitoring in this location was done by video, so we established cameras in multiple locations: the front door of the house looking into the family room, in the family room looking toward the master bedroom hallway, in the family room looking toward the kitchen and garage door, in Jessica's bedroom looking out toward the family room, in the corner of the master bedroom looking at the bed and door, on the master bedroom dresser and also looking at the bed and door, in the master bedroom and focused close-up on the inside door handle, in Melissa's bedroom looking at the frame that had moved, in Melissa's bedroom looking at the door from the far side of the room, on the kitchen countertop looking toward the family room, on the kitchen countertop looking toward the garage door, and on a smaller kitchen countertop looking into the family room. In other words, we had a view of very nearly the entire house.

Finally, we used both AC and DC (natural) EMF meters and still photography cameras on hourly sweeps of the entire house. With regard to electromagnetic fields, one concern Wayne and Charlotte had expressed to us was that there is a high voltage power line directly behind the home at a distance of about 100 feet. They worried that it could be unsafe. We measured the electric fields in the home as well as in the yard near the power lines and found no unusual or dangerously high readings. The only location in the house with higher AC EMF readings was in Jessica's room, and we attributed those reading to a large in-room cooling unit and fan that were both running at the time.

Temperatures in the house remained consistent and varied by only about 2 degrees Fahrenheit throughout the evening.

Rather than presenting a precise chronology of our investigation (because most of it was uneventful), we're going to present our findings organized by the relevant paranormal claims to which each finding applied.

The toilet flushing itself: We actually managed to solve this one even before conducting our on-site investigation. When the family first described what had been happening, we told them to perform an experiment which might sound strange but would help us understand what was happening. To test our hypothesis, we asked them to take a jar of red liquid food coloring and sprinkle a few drops into the toilet's tank (not the bowl, but the tank on the back), take a photograph of the bowl, wait half an hour, and then take another photograph of the bowl and send us both photographs.

In the before and after photographs we received from the family, some of the food coloring had flowed from the tank into the bowl. That indicates the toilet was slowly leaking. When enough water from the tank eventually dripped into the bowl, the low water level would trigger the toilet to flush by itself and start refilling. All they needed to do was replace the flapper valve on their toilet. They did so and the "ghosts" stopped fiddling with their porcelain throne.

Footsteps in the hallway: we were surprised and perhaps a little alarmed to discover that we also heard the footsteps walking down the hallway while we were investigating the house, so we did some detailed investigation of the home's construction. We discovered heat for the home was supplied by a water heating system. Even in the summer months when heat isn't needed, the system still occasionally flushes water through the pipes to keep everything clean and running.

Those pipes happen to be located beneath the house in a basement/crawlspace area. Sure enough, one of the pipes runs directly below the hallway, spanning the hall's full length. Whichever contractor either built the house or had most recently worked on the pipes was apparently lazy. The pipes had never been attached to the floor joists as they were supposed to be. The result was that whenever water flowed through, the pipes would "jump" and strike the supports. And they didn't do this all at once, but in a sort of wave pattern, moving down the entire length of the hallway.

Whenever the pipes moved, the sound could be heard upstairs in the hallway. And it really did sound to all the world like someone stomping down the hall. We told the family to anchor the pipes to the floor joists, and the footsteps in the hallway stopped.

Moving door handles: we were also able to observe the claimed movements of the door handles. It didn't look like someone was trying to turn the handles or open the door. Rather, it looked like someone was just jiggling the handle. Our seismometers found the culprit: the entire house would vibrate a little bit from time to time. Once we knew we were looking for vibrations, we found there's a large water pump located near the home, as well as some commercial equipment in nearby structures. All of these (and especially the water pump) contributed to a significant and intermittent vibration that caused the door handles to move.

Apparitions of shadow people and voices: we didn't observe any of these phenomena during our investigation. However, we suspect they're caused by a combination of the vibrations (which can sometimes cause hallucinations or uneasy feelings) and the family's own suggestibility and belief that something paranormal was happening.

Moving photo frame: this also went unobserved during our investigation. Given that Melissa couldn't be sure whether it was paranormal or whether she'd accidentally bumped the frame, we think the most parsimonious explanation is either that she bumped the frame or, perhaps, that its movement was caused by the same vibration we mentioned above.

Moving lampshade: it was a complete accident that we witnessed the lampshade moving, but we saw it tilt slightly during one of our breaks. Investigation quickly revealed its movement was caused by a loose floorboard. During the break, one of our investigators happened to step in just the right place on the offending board and it caused the lamp to tilt slightly.

Feeling of being followed while walking outside at night: This one turned out to be completely true. We discovered the phenomenon while a couple of our team members were out for a short walk to get some fresh air during a break. What you need to know about this property is that when we say it's rural, we mean it. It was pitch dark outside and there wasn't another soul around.

While walking, our people heard a growl behind them. Then more growls off to the sides. They quickly realized they were being surrounded by coyotes. Somehow, because our people have extraordinarily eclectic backgrounds, we knew exactly how to handle the situation. If you're about to be attacked by coyotes, you can scare them away using a process called "hazing." Make yourself look as big as possible, wave your arms, and yell at them. Because these are reclusive and opportunistic animals, they'll typically leave you alone and move on to what seems like it will be an easier target. Do *not*, however, run away from them. They'll take it as a sign of weakness and will likely attack.

When we returned to the house, we had to explain to Melissa and her family that her feeling of being followed while out walking the dogs at night was entirely correct. She *was* being followed, though by coyotes rather than ghosts or demons. Worse, she was a small girl who was walking a tiny dog that would easily be seen by coyotes as the equivalent of a sack lunch on a leash. The family agreed that Melissa shouldn't go out alone to walk the dog at night anymore.

Melissa's disproportionate fear of ghosts: During our interview, we were able to determine that part of the reason Melissa was more afraid of the ghosts and thought she was being individually targeted is because children can be complete buggers to each other. Her friends had convinced her that she was being harassed by the devil and murderous spirits, and her high degree of suggestibility made her the perfect target for such a prank.

In response, we helped her to realize that children play these kinds of pranks on each other and she shouldn't put much stock in what they said. We also taught her how to get back at them with some pranks of her own. And if you ever need coaching on how to construct a perfect prank, we're the guys to call. Hopefully we didn't go too far.

At the end of the investigation, we'd been able to explain pretty close to everything. Only the visible and auditory apparitions were exceptions because they didn't manifest during our investigation, but we managed to come up with at least some tentative hypotheses even for those. This is significant because of how rare it is. We're usually able to explain some things, but to be able to document and explain essentially every phenomenon reported is a rarity for us.

When we gave our report to the family, we weren't sure what to expect. Some people respond positively and are just glad to have their answers. Others become so invested in the ghost stories that they resent our results when they don't conform to their expectations. In this case, we're glad to report that the former was the case. They were not only understanding but *thrilled* to learn we'd been able to explain all the things that had been bothering them, and they set about performing some of the repairs necessary to make the phenomena stop.

They also said they'd cut back on the amount of paranormal television they watched (or at least watch it with some more skepticism).

All the paranormal phenomena stopped. They made a few repairs and adjustments to their routines, stopped frightening themselves, and it all went away. They moved on with their lives, no longer afraid of their own house. That's a big win for us.

We're also grateful that we happened to be the first group called to investigate their house. As we've mentioned before, we're not the only good team out there, but there are some groups who don't actually investigate but simply tell people their houses are haunted by ghosts—or worse, demons—as a matter of course. Had such a team preceded us to this family's home, we don't like to think what would have happened.

PART THREE
Media Analyses
and Other Activities

Activities of the Rocky Mountain Paranormal Research Society are not and have never been limited to investigations alone. While those do take the bulk of our time and represent the flagship of our work, we have our hands in a lot of other things as well. We now turn our attention to those cases that are somehow different from the standard investigations (whether public or private) we've been discussing so far.

In many ways, this third part of the book represents something of a grab bag of different and often unrelated activities, but most (not all) of them fall into a couple of categories.

First are the media analyses, and those can be subdivided into our analyses of media that has been published and received enough attention to merit our commentary and our analyses of photographs or videos sent to us along with an invitation to perform an investigation but which never develop into a full-fledged investigative activity (typically because our analysis of the photo or video is sufficient to explain the phenomenon in question).

Then there are the experiments or other forms of academic research. From time to time, independent of any specific investigation, we perform a variety of experiments whose results may illuminate some aspect of paranormal lore or phenomena. We've done several such experiments in the past, one of which will be included in this book, though our experimental "department," so to speak, is beginning to pick up more steam in recent years, so later volumes will likely contain more of those cases.

Sometimes we also get involved in cases that aren't so much paranormal investigations as they are public services or civil or political actions. Clarity is in order here. Politically, Rocky Mountain Paranormal is committed to neutrality. We're an investigative and educational organization. Members through the years have been Democrats, Republicans, and other. We don't espouse any particular political or economic philosophy, we require no political test for membership, and we're committed to remaining non-partisan in our work. However, every once in a while, a political issue emerges that is specifically related to the paranormal. When that happens, we feel compelled to take action to ensure that everyone involves has the highest quality information available and behaves ethically. None of those cases will be included in this particular volume, so we needn't offer too in-depth a commentary here, but we thought it important simply to mention that such activities have taken place and will be covered in detail in the future.

Finally, sometimes there's a case that just doesn't seem to fit into one of the other standard classification of investigations at either public venues or private residences. Even if we stretch the meaning of those phrases a bit, some of the things in which we involve ourselves just seem to belong more properly in this kind of a grab bag part of the case files books.

Though this part is relegated to the back of the book and doesn't contain as many pages as its predecessors, don't be fooled into thinking it's less important than the others. Any and every case in which we participate offers something interesting and educational, and we hope you enjoy reading about some of our less typical activities here.

ESSAY
"Grab Your Worst Camera"
A Brief History of
Paranormal Photography

People like to send us their paranormal photographs (or occasionally videos). Maybe they want us to use our experience and expertise to analyze their media and let them know what they captured. Other times perhaps they just want to show off the creepy or weird photographs they've taken. Rarely, they might be trying to perpetrate a hoax and get us to sign off on the photo's authenticity. Something we've noticed over the years, though, is that many of the photographs are simply not of very high quality.

You can probably relate. How many of us have been really excited to learn about new photographic evidence of some paranormal phenomenon only to be disappointed when the image is so blurry and out of focus you simply can't tell what it is? To an extent, it's understandable. If someone sees something unusual, it's unlikely he or she was already in the process of focusing a camera to take a perfect image. Paranormal photographs are often taken in haste, often in poor lighting conditions, and often by non-expert photographers. It's no wonder they don't always come out as nice as we'd expect.

Still, the waves after waves of terrible photography has led us to a bit of an inside joke. Whenever someone claims to see something paranormal, the joke goes, they immediately turn to their nearest friend or neighbor and exclaim: "This will completely revolutionize science. Quick! Grab your worst camera and take a picture of it." Of course, we're not saying people are deliberately taking bad photographs (with the possible exception of intentional hoaxes, but those are relatively rare), but the fact remains that the average paranormal photograph tends to look like it was taken with a potato instead of a high-end professional camera. This, despite the modern ubiquity of cameras whose quality would have shamed even the most expensive professional devices of only a few years ago.

When photography first entered the mainstream, it wasn't a hobby just anyone could pick up. It required a lot of technical knowledge and expensive equipment. In fact, it's largely due to the Stanley brothers that it became the widespread hobby it is today (see Chapter 14). Almost as soon as photography became accessible to the public, it was being used to capture photographs of spirits (and before long a wide variety of other paranormal entities). It's generally thought that American amateur photographer William Mumler produced the first photograph of a spirit some time in the 1860s. Mumler's image represents not only one of the first "viral" photographs that received a lot of attention (and even praise from so-called experts who testified to its authenticity), but one of the first deliberate photographic hoaxes created by means of a double exposure.

When Mumler was eventually put on trial for fraud, no less a historical giant than P. T. Barnum testified against him in court, producing his own fraudulent spirit photograph (created by Abraham Bogardus, who Barnum had hired for the purpose) of President Abraham Lincoln to demonstrate how easily a fake could be created. He also pointed out that many of the "spirits" in Mumler's photography turned out to still be among the living. Arguably, the Barnum/Bogardus photograph was of higher quality even than Mumler's, but the jury wasn't convinced and Mumler was ultimately acquitted. It remains an open question whether, had he been found guilty, the history of spirit photography might have been stifled. Probably not, but we'll never know.

As photography grew in popularity and entered the mainstream throughout the late 1800s, interest in spiritualism was very much on the rise, and countless photographers, both amateur and professional, developed a variety of techniques for fraudulently creating spirit photographs. Some of these were deliberate hoaxes intended to get the photographs published. Others were deliberate hoaxes meant to provide further evidence in favor of a psychic medium's abilities (and to draw paying customers to seances, which were also growing in popularity at the time). Still others were never intended to defraud, but were the kind of harmless amusements lots of us like to tinker with in our free time. Indeed, even to this day, people like to create spirit photographs not to claim they're real but just for the art of it. Members of Rocky Mountain Paranormal have created more than a few of our own, both for experiments and just for the fun of it. Some of our members who are trained in the skills of magicians or conjurers have also occasionally been known to produce spirit photographs on *another* individual's camera during a live performance, something the early spirit photographers could never have dreamt of.

Something was different about the early spirit photographs, though, when compared to those offered today. They tended to be of higher quality. Photographic plates were expensive and photographs were not nearly so disposable as they are today. In those early days, one only took a photograph when one was really sure it was going to be a good one. They didn't take hundreds in the hopes that one of them might turn out like many amateur photographers (and even some professionals) do today. As such, they put a lot of effort into their hoaxes.

And hoaxes they were. The advent of fundamentally honest people legitimately believing they captured something paranormal on film hadn't yet occurred. These early photographs were cleverly doctored and made to look to all the world like they contained real spirits. Things like "orbs" and other nebulous paranormal entities hadn't yet entered the game.

To maintain our neutrality, we have to maintain the possibility that maybe some of the spirit photographs of the time were the genuine article. However, we can say with certainty that at least most of them were frauds, which we know because the frauds were eventually discovered by diligent and skeptical investigators. We can also say that if any of them are genuine, no scholars have yet identified any kind of tell-tale hallmark that would separate the real from the fake.

It's also worth mentioning that, though we've specifically mentioned the double exposure as a method of faking a spirit photograph, other ideas were also developed. Some involved modifying cameras, tinkering with the photographic plates (later film), or doctoring photographs after they'd been taken.

Spirits, similarly, weren't the only subjects of paranormal photography. One infamous incident occurred in 1917 when two young girls—Frances Griffiths and her cousin Elsie Wright—produced a series of photographs depicting themselves interacting not with ghosts but with fairies. The Cottingley Fairies, as they came to be known because the photographs were taken in Cottingley Glen, took the world by storm. No less a literary giant than Sir Arthur Conan Doyle (creator of Sherlock Holmes) wrote about his unshaking belief in the photographs' legitimacy.

Unfortunately for Sir Arthur, they, too, were fakes. The two girls had cut out pictures of fairies from a book and placed them in the photographs they took. Initially, it was meant to just be a harmless prank, though eventually the story got away from its creators' control and developed into a substantial controversy which is still discussed more than a century later.[1]

As cameras and film became cheaper, more people began taking photographs, and the world of paranormal photography entered a golden age that persists to this day. But it's not the same as it was. Rather than intricate intentional hoaxes (though there still are a few of those), the "fake" paranormal photographs today and throughout recent decades have mostly been honest mistakes made by people who were carelessly

---

1   Arguably the definitive treatment of the subject can be found in: Randi, J. (1982). *Flim-Flam! Psychics, ESP, Unicorns and other Delusions*. Amherst, NY: Prometheus Books.

taking photographs (why not—especially now that cameras are digital, it costs nothing to take a bunch of extra shots), often without being properly trained in the operation of their cameras.

Images containing phenomena like "orbs" or other nebulous shapes will be discussed in greater detail in the next chapter, but to explain the general idea in advance, they're the result of optical anomalies that may not be fully understood by the photographers.

Other photographs or videos containing blurred or out of focus images also leave a lot up to the viewer's own interpretation, and these low-quality photographs dominate the recent history of paranormal photography.

Why? After all, as technology improves, the overall quality of photographs people are taking seem to be getting better rather than worse. It's now possible for a complete amateur with no training at all to take a photograph whose quality (if not artistic composition) may rival that of any professional. And that's just using a smart phone camera, much less an expensive photography setup. Shouldn't paranormal photography be improving along with the technology instead of staying stagnant?

We suspect that a large part of the reason paranormal images are still of low quality is that it's those low resolution images that can be mistaken for a paranormal entity even if one wasn't present when the photograph was taken. Ambiguous images allow us the freedom of interpretation to see whatever we want in the photographs. High quality images rarely (not never, but certainly rarely) show anything that can be interpreted as paranormal.

Human psychology also plays a role. Given ambiguous stimuli, our brains are wired in such a way that we immediately try to find a signal within the noise, to make sense of what we're seeing. It's true of all of our senses, but particularly true of vision. And it's true of any pattern, though we have a particular affinity for faces. This is a phenomenon called pareidolia. If you stare into random visual stimuli, you're likely to be able to pick out something that looks recognizable (often as a face). It's why we're able to make out familiar shapes in clouds as well as (we would argue) why some people think they see religious icons in cheese sandwiches. Indeed, even something so simple as a colon and a parenthesis can be recognized as a smiling face.:)

Add to that, that in addition to the boom in high quality photographs, the widespread availability of cameras has also multiplied the number of, frankly, crappy photos, and we get ourselves to the present day, in which everyone is a would-be paranormal journalist and there are simply so many photographs allegedly depicting paranormal phenomena out there that it's all but impossible to separate the fakes from the genuine article (if, indeed, the genuine article exists).

That doesn't stop us from trying. We spend much of our lives taking photographs and videos in allegedly haunted locations, hoping to find some anomalies that might provide some evidence of something paranormal or otherworldly. Occasionally we find something that, while it may not be proof positive, at least raises some eyebrows. Most of the time we don't. The important difference between our work and the bulk of paranormal photography circulating on the Internet, though, is that we're highly trained in both photography and photo analysis so we can avoid user errors or optical anomalies most of the time (and recognize them for what they are when they do occur), so when we get something weird, most experts are more inclined to pay attention than if some unknown person sends them a random photograph. Furthermore, we carefully document the conditions under which our photographs are taken so we can at least attempt to account for the anomalies.

24
The Photographic Orbs

By far, the most common type of paranormal photograph depicts what has come to be called an "orb." These are images in which circular or spherical objects are seen, apparently glowing, even though the people who took the photographs never saw the objects in real life.

Figure 24.1. An example "orb" photograph. In the original image, the anomalous spherical shape was blue in color. Photo: Bryan Bonner.

These orbs, along with other photographic anomalies like "vortices" (usually spelled "vortexes," but we refuse to incorrectly pluralize a word) come in a variety of shapes and sizes and can actually be caused by a few different optical phenomena. A detailed and more technical explanation should appear in a future book we have planned that will deal explicitly with the more technical aspects of paranormal investigation, but for now, a brief lesson is necessary as the background to the experiment we'll discuss in this chapter.

The culprit, put simply, is usually dust or some other kind of particulate matter in the air. It may be so small you can't see it with the naked eye, but cameras and eyes don't work exactly the same way. Despite protestations to the contrary, there's always a certain amount of dust or particulate matter in the air, unless you're taking a photograph in a very expensive clean room (and even then, there's a margin of error). You can prove it to yourself by looking at a beam of light in a darkened room (for instance by looking at the light coming from the projector in a movie theater). You will see bits of dust and other "bright specks" floating around in the light.

The problem isn't the dust alone. It's the presence of particulate matter in conjunction with the camera's flash. Cameras won't necessarily "see" the dust just as we don't notice it in day to day life. However, if the dust particle is directly in front of the lens and brightly illuminated by the flash, the light can bounce off the dust and reflect into the camera's sensor, producing the appearance of an orb in the resulting photograph.

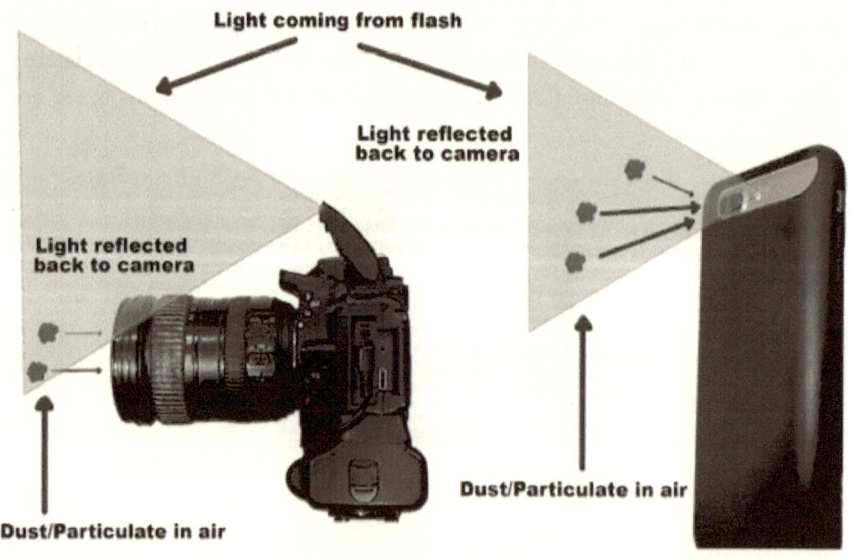

Figure 24.2. Optical explanation of orb phenomena. Photo: Bryan Bonner.

And it's not just dust. Rain, fog, mist, or other weather conditions can produce the same effect. It's also possible to capture a reflection off the condensation of your own breath if you happen to breathe directly in front of a camera with a flash (particularly on a cold day, though there's always some moisture in breath, so it *can* occur even in warm weather).

To illustrate, we present three examples we know not to be paranormal in nature because we took them ourselves and can definitively attest to the precise conditions that created each photograph. The first (shown at the beginning of this chapter) is a traditional orb. It's the result of a particle of dust in front of the camera, which we know to be the case because we shook a dusty pillow in the room before taking the photograph.

The second is a sort of fog or mist, which we think looks quite creepy. Had this photo been taken in a place like a haunted cemetery, we could easily see it being blamed on a ghost. Instead, we know it to be condensation from the photographer's breath. Before snapping this picture, one of our people exhaled deeply in the direction of the camera's lens. Incidentally, it was only slightly cold out that day.

Figure 24.3. Anomalous "mist" photograph caused by flash reflecting off the photographer's condensed breath. Photo: Bryan Bonner.

Finally, we present a photo of what is generally called a vortex (though we maintain it's a misnomer, as "vortex" implies a kind of circular motion). In this case, the culprit was the camera strap dangling in front of the lens when the photograph was taken.

Figure 24.4. Anomalous "vortex" photograph caused by camera strap in front of the lens. Photo: Bryan Bonner.

Other examples have included photographers accidentally capturing images of their own fingers, their own hair, or (very commonly) insects flying in front of the camera.

That leads us to the Sony Orbs experiment. We wanted to deepen our knowledge of the photographic processes that produce these images as well as to try to explain why orbs seem to take three distinct shapes: circular, half circular, and diamond.

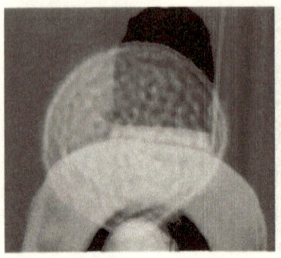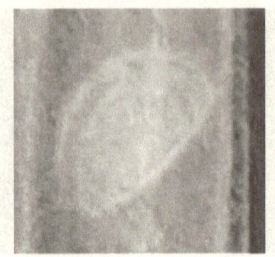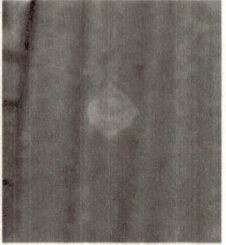

Figure 24.5. Circular, half circular, and diamond type orbs, respectively. Photo: RMP archives.

We knew that the orbs were the result of the optical properties of the camera but weren't sure why cameras of the same type would produce different sizes, shapes, or textures of orbs. To deepen our understanding, we performed a simple experiment using two Sony HD camcorders and a Nikon D70.

First, we took photos at the same time and location using all three cameras without triggering the flash. None of the three images contained any orbs, as expected given our optical theory of orb production.

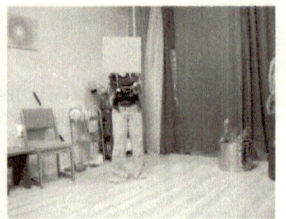

Figure 24.6. "Orbless" photographs taken without flash. Photo: RMP archives.

Next, we took photographs with all three cameras, again at the same time and place, but with the flash enabled on one Sony camera and the Nikon camera, but disabled on the second Sony. For the Sony camera, we used the built-in flash, which is located directly above the lens. For the Nikon camera, we used a flash mounted on a bracket which held the flash bulb several inches from the lens. Only the Sony camera with the flash enabled produced orbs. Had these been paranormal phenomena rather than an optical artifact, all three cameras should have shown orbs in the same location.

Figure 24.7. Left: Sony photograph with no flash and no orbs. Center: Sony photograph with flash and several orbs (circled). Right: Nikon photograph with flash mounted away from lens and no orbs. Photos: RMP archives.

Additionally, we noted that when we took photos of the same subject at the very same moment with the two Sony cameras (both with the flash enabled), orbs appear in both images but at different locations.

Figure 24.8. Two photos taken at the same instant using two identical Sony cameras. Both had flashes enabled and both produced orb anomalies (circled). Note that the orbs are in different locations. Photos: RMP archives.

Close examination of the photographs obtained during the experiment revealed that different sizes, shapes, and textures of the orbs were a function of the distance between the camera lens and whatever particulate matter caused the orbs. Orbs near the center of the field of view were more likely to be circular, while those at the periphery were more likely to take on the shape of the rounded edges of the lens and lens hood.

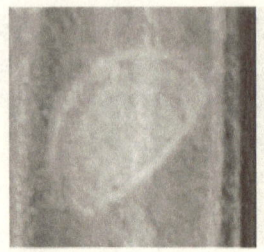 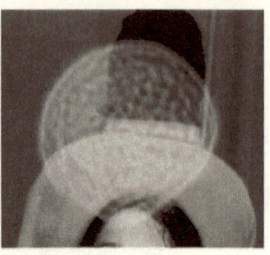

Figure 24.9. Left: orbs occurring at the periphery of a camera's field of view took the shape of the lens and lens hood, resulting in the "half circular" type orbs. Right: orbs near the center were of circular type. Photos: RMP archives.

Some of the orbs took on a diamond shape, which was a bit of a mystery for a while. We reached out to Sony and discovered the iris/diaphragm inside the lens on these cameras was, in fact, diamond shaped. Depending on the depth of focus in the image, this produced the diamond-shaped orbs.

Why should anyone be interested in this experiment? It's relevant to paranormal investigation because it shows that not only can orbs be produced by particulate matter, but a variety of factors can affect their size, shape, and texture.

People often send us their orb photographs for analysis. It's a known phenomenon that dust can cause orbs, so we've noticed that a lot of the letters accompanying those images claim that while they acknowledge orbs can be caused by dust, *these particular* orbs couldn't be. How they made that determination is never specified, but we suspect they're being misled by the morphological diversity of the orbs in their photographs. They may think dust produces only one size or shape of orb image.

In reality, though, as our experiment shows, it's not the nature of the dust but the precise conditions under which a photograph is taken—and in particular the relationship between the lens, the flash, and the particulate matter—that determines whether or not orbs will appear and, if so, what they'll look like.

25

The 1971 Costa Rica UFO Case

As you should be able to guess from the title of this chapter, our subject is a UFO that was witnessed in Costa Rica in the year 1971. That predates the founding of the Rocky Mountain Paranormal Research Society, and any reasonable classification would have to consider it something of a cold case. Paranormal cold cases are something we do take an interest in, though they don't form the bulk of our work (despite the overwhelming number of such cases available) for a variety of reasons.

First, by the time a case is a few years old, there's a good chance other organizations (believers and skeptics alike) have already analyzed it to death. Every once in a while, if there's a compelling public interest in the case, we might wade into it anyway, but as a general rule, if someone's already done a satisfactory analysis we feel like our efforts would be better spent elsewhere.

More importantly, cold cases are simply difficult to investigate. People often approach us after our lectures and share their ghost or alien stories and ask how we might be able to explain their experience. Simply: we can't. We weren't there. We couldn't observe, document, measure, test or in any way begin to perform a forensic analysis. All we have to go by is the individual's own story. Most of the time we don't think they're lying, but there are all kinds of reasons why their recollection of the experience may or may not be reliable. There's just nothing we can do except speculate, and if we were to put a case in our case files, we'd want to be able to do at least something more than mere speculation.

Sometimes, though, a cold case presents itself in such a way that these negatives aren't as big a problem. We will attempt to tackle cold cases if there's sufficient interest in the case to merit our attention, few or no other forensic analyses have been performed or published, and there is sufficient evidence on which we can base our investigation. This is such a case.

On September 4, 1971, the Costa Rican National Geographic institute was conducting an aerial mapping mission. While flying over a location known as Lago de Cote, photographer Sergio Loaiza captured an anomalous image which, in addition to the coast line they were attempting to photograph, depicts what appears to be a disc-shaped object in the corner of the photograph.[1] In the years (now decades) since, this has become regarded by many as one of the best and most convincing UFO images ever recorded.[2]

We pause for a moment to emphasize the first letter in that initialism. "UFO" stands for "Unidentified Flying Object." Though UFO cases are connected with extraterrestrial lore, it's important to recognize that just because something is unidentified does not automatically make it alien. In fact, once you start calling it an alien, you've identified the object, making it, we suppose, more of an "IFO" (an Identified Flying Object). Still, despite our terminological issues, there's certainly an interest in trying to determine what these objects are. If they are indeed extraterrestrial in origin, that would change everything. And if not, there may be national security reasons we should want to know what's going on in the skies (which is why governments have, without admitting the existence of extraterrestrial beings, taken a keen interest in the study of these phenomena).

For several years, the image was just a curiosity. But in 1989, it was analyzed by two individuals—Richard F. Haines and Jacques F. Vallee—who concluded the anomalous object in the photograph was disc-shaped and had "a maximum dimension" of 683 feet.[3] Obviously that's a large object, so if their analysis is correct, we *certainly* need to know what this thing actually is. After all, the world's largest airplane, the Antonov AN-225, had a length of only 276 feet.[4] That's less than half the alleged size of our

---

1   Because the photograph in question is still under copyright, we have chosen to omit it from this publication out of an abundance of caution. It can easily be found online following a web search for "Costa Rican UFO photo" or (as of this writing) embedded in the *Medium* article located at <https://medium.com/on-the-trail-of-the-saucers/costa-rica-ufo-photo-c0b1eb07c5e7>.

2   For example: Zabel, B. (2021). The Best UFO Photo Ever Taken? *Medium.* <https://medium.com/on-the-trail-of-the-saucers/costa-rica-ufo-photo-c0b1eb07c5e7>

3   Haines, R. F., & Vallee, J. F. (1989). Photo Analysis of an Aerial Disc Over Costa Rica. *Journal of Scientific Exploration, 3*(2): 113-131.

4   Only one AN-225 was ever produced. It was made during the Cold war by the Kiev-based Antonov company. The plane, nicknamed "Mriya" (Ukrainian for "dream"), was destroyed by Russian forces during Russia's 2022 invasion of Ukraine.

unknown object. Even the Saturn V rocket that carried men to the Moon was only 383 feet tall. This thing, if the Haines/Vallee analysis is correct, is *huge*.

So who are these men who performed the analysis? Dr. Richard F. Haines, Ph.D. holds a doctorate in the field of experimental psychology. For 19 years, he worked for NASA-Ames, conducting research for displays, window design, and space suit habitability. He founded the Joint American-Soviet Aerial Anomaly Federation to study, essentially, UFO cases, on which he has published numerous papers.

Jacques Fabrice Vallee holds a B.S. in mathematics and an M.S. in Astrophysics and co-developed the computerized mapping of Mars for NASA. He, too, is a UFO enthusiast, and has defended the scientific legitimacy of the extraterrestrial hypothesis and even an interdimensional hypothesis. Additionally, he's been involved in numerous projects for computer network applications and is on the scientific advisory board for Bigelow Aerospace, a private aeronautics company.

All of this is to say, these men do have some legitimate credentials. And though they certainly have their biases as UFO enthusiasts, we also agree with them that these phenomena should be the subject of proper and open-minded scientific analysis. However, we do note that their credentials as a psychologist and a computer scientist are not necessarily aligned with the task of forensic photographic analysis. That doesn't make their analysis wrong, by any means—amateurs make many legitimate contributions to many fields—but it's something to just keep in mind.

Their analysis was commissioned by the Society for Scientific Exploration, a "professional forum for presentations, criticism, and debate concerning topics, which for various reasons, are ignored or studied inadequately within mainstream science," according to their mission statement.[5] Further, their research covers "subjects ranging across consciousness research, progress in new energy technologies, discoveries bearing upon healing and medicine, and more."[6]

Now that we know the players in this drama, let's take a deep look at the photograph itself. It was taken using a Carl Zeiss RMK 15/23 camera system which was set to capture an image once every 17 seconds.

According to Haines and Vallee, "if the disc was located 10,000 feet away from the camera, its maximum dimension would be 210 meters (683 feet)."[7] They also explained that "the absence of a shadow from the disk remains a puzzle to us…an obvious explanation is that the object is at the surface of the Earth where no shadow would be expected. Another explanation is that the object is opaque, small, and much nearer the airplane so that its shadow's reduced size and darkness would be difficult

5   SSE (n.d.). About the SEE. *Society for Scientific Exploration.* <https://www.scientificexploration.org/about>

6   *Ibid.*

7   Haines, R. F., & Vallee, J. F. (1989). Photo Analysis of an Aerial Disc Over Costa Rica. *Journal of Scientific Exploration, 3*(2): 113-131.

or impossible to locate on the ground."[8]

We have some questions, to say the least. The authors seem to give preference to the assumption that the object is on the order of 10,000 feet from the camera and use that assumption to ascertain its size based on its apparent size in the photograph. They gave only cursory consideration to the possibility that the object could be smaller and closer to the camera. Indeed, if that were the case (particularly if it were very small and very close to the camera), no shadow would be expected, so that seems like a more parsimonious explanation.

The problem is further exacerbated by the lack of witnesses. A report concerning the image issued by the Costa Rican government (whose people had captured the photograph), said that when the image was taken, "none of the flight crew or photographers saw the object."[9] Given they were involved in a mapping mission, all of the crew and photographers would have been keeping their eyes outside the plane. Had something larger than the world's largest aircraft shown up directly below them, it seems like they should have seen it.

Haines and Vallee did include a section regarding sightings of the alleged object from the ground, but they include sightings that occurred on dates *other* than the date the photograph was taken, including September 27, 1978 and October 25, 1986. While it's possible these sightings could be legitimate and even possible they could be of the same object, there's no way to correlate them with an event in 1971 without some other kind of evidence to connect them. No such evidence has ever been presented.

Because none of the crew saw any anomalous object, we had to consider the possibility that an optical phenomenon was at work rather than a large (potentially extraterrestrial) object. One idea we considered was that some small object happened to pass in front of the camera at exactly the right moment, but it would have been difficult to identify such an object so we turned our attention first to the camera itself, which was never considered in the prior analysis.

As soon as we started researching, we noticed that the camera has a rather unusual optical system, the Carl Zeiss RMK 15/23 optical system, which looks a lot like the object shown in the photograph. We thus believe the image was created by reflections of ambient light inside the optics of the camera system. Therefore, the image does not likely depict a flying object but is probably the result of a combination of the type of camera used, angle of the plane to the light, position of the sun, and possibly the angle of light reflecting from the surface of water below the plane.

We would like to get our hands on one of these camera systems and an airplane and attempt to recreate the image. However, our funding limits our ability to do so. Still, we consider our explanation to be the most parsimonious. It also explains all elements of the image and circumstances surrounding its capture, which the alternative

8   *Ibid.*
9   *Ibid.*

explanation did not—recall that the Haines/Vallee analysis was unable to account for the lack of a shadow from the object, whereas our analysis relies on phenomena within the camera which would not cast a shadow.

26
The Lafayette UFO

Ahh, Lafayette. It's a great city, and it's home not only to some ghost stories (as is pretty much every city in the world) but even a vampire (see Chapter 8) and now a UFO story that obtained more than its share of media attention a few years ago. Unlike the case outlined in the previous chapter, we became involved in the Lafayette UFO story even as it was developing.

Unfortunately, this represents one of those cases for which available information is limited, and this will accordingly be a short chapter.

In late March of 2011, multiple individuals in Lafayette, Colorado reported seeing UFOs (in the true sense of the word—they could not identify what they'd seen) in the evening sky. The anomaly took the form of three lights arranged in triangular formation. Video of the event depicts the lights as non-moving or slow-moving points of light without any visible structure (this is certainly partially due to the fact the video was filmed at night in low-light conditions and may also be partly due to the distance between the camera and the objects, though the distance was not easy to judge). The lights' color appeared, at least on video, to be a reddish orange hue, though it's similarly difficult to determine whether that's exactly how they appeared in person or whether some of their coloration was an artifact supplied by the camera.

Especially because more than one person witnessed the phenomenon, the media took a keen interest in the case (though it was quickly forgotten in favor of more pressing news of the day). Reporters reached out to local airports as well as the Federal Aviation Administration (FAA) and were told that nothing out of the ordinary had been picked up on radar, though officials didn't offer any further comment.[1]

We began looking into the case at the same time as MUFON, the Mutual UFO Network, did. Over the years, we've worked with MUFON in an unofficial capacity on a few projects and our paths have crossed with them numerous times. They're essentially in the same business we are, albeit focused specifically on UFOs while we look into anything strange or bizarre. Probably the MUFON people think we're a little too skeptical and we tend to find them a little too credulous, but at least in principle our philosophies are much more aligned than they are different.

The problem was, for both our team and the MUFON team, the video simply wasn't of high enough quality to perform a detailed analysis. It depicted lights in the sky. On that we could all agree. But without other background images to use as reference and given the film's low resolution, we weren't able to determine much about the object or objects. One of the first things we always try to determine is how far from the camera an anomaly was. That will help us determine things like its size and speed. But in this case, our guess was as good as anybody's.

We do believe the anomaly was not merely an optical phenomenon. It was reported by multiple witnesses who claimed to see it with the naked eye, so we know it wasn't just the product of a particular photographer's error or a malfunctioning camera.

About a week after the sighting, we were invited to discuss the case on the local ABC News affiliate. Also present at the discussion was a representative from MUFON. Despite approaching the case from slightly different angles, we weren't able to give them the spirited debate they might have been looking for. Both representatives—ours and MUFON's—emphasized the importance of ruling out any possible natural or mundane explanation before reaching for a supernatural or extraterrestrial one. Neither organization had yet been able to rule out such explanations.

During the interview, we mentioned that the region had been experiencing a number of wildfires at the time, which prompted a certain degree of usual aviation activity—slurry bombers, helicopters, etc. Though we were unable to confirm those kinds of aircraft were what people had seen, they were and remain what we consider to be the most plausible explanation.

After the interview, we wanted to see if we could take our investigation any further. We still weren't able to find any official flight records or confirmation that what people saw had been aircraft. But we were able to figure out that the apparently slow speed of the craft would not have been anomalous.

We reached out to a pilot of our acquaintance and asked, given the load a plane would bear if it were engaged in firefighting operations (which would slow the plane down) and the high wind conditions during the incident, how fast a plane might actually

1   Kuta, S. (2011). Some baffled after UFO sighting claims in Lafayette. *Daily Camera*. <https://www.dailycamera.com/2011/03/24/some-baffled-after-ufo-sighting-claims-in-lafayette/>

be moving. Obviously it has to move to stay in the air, though its motion only needs to be relative to the wind, so if it were flying directly into a strong wind, the pilot told us it might not appear, relative to the ground, to be going any faster than 25 or 30 miles per hour. Because we knew that many such overloaded planes were in the air as part of firefighting operations, we're quite confident (albeit not quite 100% certain) that we've figured out what the "Lafayette UFO" actually was.

Not long after this case entered the public attention with much fanfare and then fizzled out for lack of investigable evidence, we were party to another UFO case that did lead to more definitive conclusions, but for that you'll have to wait for a future volume of this series.

27
Puckett's Ghost: The Case of the Junkyard Ghost

Most of our cases are opened and closed fairly quietly, with relatively little fanfare or media attention. Though we've become some of the go-to experts on the weird for local (as well as national) media, most of our work involves cases that either receive no media coverage to speak of, or which receive only local mention. And then there's this case. Even before we got involved, it took the national media by storm, and has been featured in any number of television shows and publications. In fact, it was because of its media prominence that we became involved, and it ended up being quite an interesting little case.

One evening in September of 2002, the staff on duty at Puckett's Wrecking Service, an auto yard in Oklahoma City, Oklahoma, noticed something strange on their surveillance footage. What appeared to be a ghost entered the frame, moved around for a bit, and then exited the camera's view. They saved the surveillance footage, and most of the story that follows centers directly on what the camera captured. If you want to see the full video, it's readily available online[1], or you can approach us at one of our public lectures and we'd be glad to call it up for you.

That same evening at around 2:00 a.m., two staff members were working at the yard and one of them went to search the lot after seeing an image on the monitor. No one was there and they didn't see anything out of place.

Both the owners and staff members of the wrecking yard thought they'd discovered something fairly interesting (and they were even more correct than we think they

1   https://puckettswrecker.com/gallery/

likely suspected at first) and took their story and footage to the local press. A newspaper ran the story, complete with a description of the video and commentary from workers who by this point were theorizing that what they'd seen may have been the ghost of a woman involved in a deadly automobile accident. After all, the wrecking yard was full of cars that had been involved in accidents, and some of them were lethal.

The first group to tackle the Puckett's Ghost was G.H.O.U.L.I. (Ghost Haunts of Oklahoma and Urban Legends Investigations).[2] Their analysis concluded that the video could not have been faked without a large special effects budget. Other paranormal researchers also jumped into the fray to offer their opinion that the tape was an authentic video of a ghost because it would have been too difficult to fake.

Over the years, we've heard similar claims regarding a wide variety of paranormal video tapes. Everyone knows it's possible to do essentially anything with Hollywood-style special effects, so a common refrain is that the particular video in question would have been prohibitively expensive to produce. We don't know for certain, but our strong suspicion is that these pronouncements are made without consulting film or television special effects professionals.

In fact, we suspected such was the case here, so we reached out to Puckett's to see if we could get a copy of the recording. They complied and we set to work on our analysis. Upon reviewing the film closely, we determined that not only was the recording from a low-resolution security camera, it was located outside of the auto lot's boundaries on a light pole. The relevant portion of the video lasts 29 seconds.

Scanning through the image frame by frame and watching it repeatedly at a variety of different speeds, we noticed a few additional things. First, when the image of the ghost appears within the camera's view, it seems to "drop down" into the image. At the end of the film, the ghost appears to move off camera rather than to disappear. That certainly suggests someone could have simply moved something in front of the camera to create the illusion of a ghost, though it doesn't prove it definitively.

What would prove the matter conclusively would be if we were to find a "smoking gun" in the form of a support wire being used to hold a doll in front of the camera. A couple frames of the video contain anomalies that hint at this possibility, but the film is of such low resolution that we can't definitively separate a potential string or wire from "noise" in the image. Still, we suspected that dangling a doll of some sort in front of the camera was likely the method by which a hoaxer may have produced this video.

Because the camera was mounted on a light pole, we tried to think of different ways the ghost could be placed in front of the lens and where the support line could be attached. One idea we thought of was an old stage illusion known as Pepper's Ghost.

Invented in 1862 by English scientist John Henry Pepper, Pepper's Ghost is an illusion originally used on stage but later adapted to not only theatrical productions but

---

2   Not to be confused with "Ghouli," the fifth episode of season eleven of *The X-Files*.

film, amusement parks, magic shows, and even haunted houses. It's a means of creating the appearance of a ghostly or semi-transparent apparition on stage to interact with other actors or in a location where such an entity ought not to be. It's created by reflecting a brightly lit object (which will become the image of the ghost) onto a piece of glass placed at a 45-degree angle to the plane of the viewer.

Figure 27.1. Illustration of Pepper's Ghost published in Le Monde Illustré, 1862. Public domain.

When this is done well, the brightly lit object will appear semi-transparent and quite ghostly. It can be done at any scale depending on one's budget (large sheets of glass and the frames to hold them can become pricey), and the "plane of the viewer" can refer to an individual watching in a live performance or a stationary camera.

Most famously, Pepper's Ghost is used in the Haunted Mansion attraction at the Disneyland and Disney World amusement parks. Visitors to this attraction are able to see an entire ballroom filled with semi-transparent ghostly figures. This is accomplished by brightly lighting animatronic "ghosts" and reflecting the image onto a sheet of glass just as in Pepper's design.

With regard to Puckett's Ghost, we were able to recreate the junkyard video using a Pepper's Ghost apparatus. We used a Radio Shack low light camera with roughly the same resolution as the surveillance camera. For the ghost, we used a 5-inch G.I. Joe action figure. We attached it to some black string. We recreated the lamp atop the light pole using a flashlight, and we used a 2-foot by 3-foot sheet of clear acrylic for the Pepper's Ghost glass.

Our video looked remarkably similar to the Puckett's Ghost video. Obviously the particular shape of the character was different as we'd used a G.I. Joe and we don't know what kind of doll the original might have employed. But the images look quite similar, especially when viewed on video instead of still images: the movements of the two "ghosts" are nearly identical.

Figure 27.2. A frame from our Junkyard Ghost recreation video. Photo: RMP archives.

We reached out to Puckett's and shared our findings, explaining that we thought the video was a hoax. Importantly, we didn't suggest *they* were the hoaxers. The light pole in question is outside of their property, so just about anyone with a little know-how and some spare time could have done this. But they'd hear none of it and wanted nothing more to do with us.

Not long after, National Geographic got involved. They were working with Puckett's to feature the story on an episode of a new television series called *Is It Real?*[3] Much to the chagrin of many in the paranormal world, they interviewed us for the program, and we also re-created our version of the video for their audience. We were pleased with the way the show handled the situation. Their goal, and we think they were successful in it, was to provide an objective look at a variety of paranormal phenomena.

After going to all the trouble of using Pepper's Ghost for our recreation, we eventually discovered it wasn't necessary. The video was of low enough resolution that it was possible to re-create the video simply by dangling a figure in front of the camera. This could have been accomplished simply by attaching a doll to a string dangling from a long rod or pole and moving it in front of the camera. This could also account for the ghost's erratic movements in the video.

We're not sure which method the hoaxers employed, but we're convinced that either method is distinctly possible. Other skeptics have suggested a non-deceptive

3   Cascio, M., & Conboy, M. (executive producers) (2005). Ghosts. [TV series episode]. In *Is It Real?* National Geographic.

explanation. Namely, they think the video depicts an insect hovering near the camera's lens.[4] We don't agree. While it is absolutely the case that insects near camera lenses have resulted in any number of alleged paranormal photographs or videos, the Puckett's video contains two elements which make us suspect a deliberate hoax rather than a mistaken shot of an insect. First, the alleged ghost's motions follow a roughly circular pattern which is consistent with an object dangling from a string. Second, though the image is blurry, it appears to bear greater resemblance to a rough humanoid rather than insectoid shape. Put simply, we've seen a lot of insect videos and a lot of hoaxes over the years, and this one looks a lot more like the latter.

As hoaxes go, though, it's a pretty good one. We've added it to our own toolkit in case we ever need it when we're filming a low-budget horror movie.

---

4   Nickell, J. (2013). Junkyard Ghost. *Center for Inquiry Investigative Briefs*. <https://centerforinquiry.org/blog/junkyard_ghost/>

Afterword

Our case files represent nearly a quarter-century of hard work. And we're still at it and going stronger than ever. In that time, we've seen cases that amused us, baffled us, made us laugh, made us angry, and everything in between. There's a temptation to say "we've seen it all," but the fact of the matter is, we've only just scratched the surface. The world is a very big and very weird place. With apologies to Hunter S. Thompson, "when the going gets weird, the weird turn pro."

This book doesn't even come close to covering everything we've experienced so far. It's only the first volume in what will already be a multi-volume series. And with a little luck, we can make it an *ongoing* series as we keep investigating more and more of the strangest things the world has to offer.

Each of these investigations has been an adventure. Writing them into this book has been likewise. And for us, that's a big part of what this is all about. Science and history are both important and we try to do our best to educate people in those fields, using the ghost stories as a framework to do so. Ghost stories themselves, along with all the other claims of aliens, cryptids, vampires and everything else, are also fascinating. No one would do what we do, much less stick with it for as long as we have, if they weren't passionate about the subject.

So now that we come to the end of this first volume of our case files, the question becomes: are we trying to get people interested in the ghost stories or are we trying to debunk them? It's both and neither all at the same time. Hopefully our explanations give people some food for thought and help people to think about bizarre claims a little more clearly and a little more critically. But when it comes down to whether or not we're suggesting you (or anyone) should believe in the paranormal, the answer is: we don't really care. We think the stories are worth telling whether or not you believe in them. At the same time, we think they should be subject to the strictest of skeptical scrutiny whether or not you believe in them. If we accomplish both of those things—preserving as well as critically examining our folklore—we're completely happy and it doesn't matter in the slightest whether one skews more to the "believe" or "disbelieve" side of things.

What about us? Do we believe in the paranormal? Even that's a hard question to answer. Now that you've read the first volume of our case files, you've seen us find plenty of rational explanations for plenty of cases. On the other hand, you've also seen us try our best to find those explanations only to be left scratching our heads. Future volumes in this series will reveal a similar pattern. All we can really say is that we know not every claim of the paranormal is true. Many of them are hoaxes or can be naturally explained. But as for the rest? We're pretty sure at least some of them will turn out to be completely mundane. Could all of them, though? It's certainly possible, but it's not a guarantee.

The way we see it, the world is full of mysteries. Every time we investigate something, there's an opportunity for discovery. An opportunity to unlock one of those mysteries. Maybe the solution will turn out to be the ghost/vampire/alien/whatever or maybe it will turn out to be something none of us ever even thought of. It would be the height of arrogance to assume that our present level of science has fundamentally uncovered everything there is to the world, and while our relative confidence that any particular paranormal hypothesis is true may be low, our confidence that "some kind of something" weird actually is out there is much higher. The only way we'll be able to find it, if indeed there is anything to find, is to keep looking. Even if we don't find what we're looking for, we're almost guaranteed to learn something new and have a great adventure.

Until next time, then, thanks for reading. Now go tell your friends a good ghost story.

If you've enjoyed this book, we'd appreciate it greatly if you were to leave us a review online. Even a review of just a few words really helps us to get the word out.

Acknowledgements

Though writing a book is often seen as a solitary endeavor, it necessarily involves the help and support of numerous people. That's doubly the case when the book describes real-world case files spanning a quarter of a century. As such, the number of people to whom we owe our sincere thanks is beyond our ability to count, and any attempt at a complete listing would be doomed to failure. Nevertheless, we'd like to begin my offering our sincere thanks to all of those who've allowed us to conduct our investigations in their homes or businesses over the years. Without them, this book clearly would never have happened. Similarly, we'd like to thank everyone who has joined us on investigations over the years. Both groups of people would require another full-length book to list, so we thank them all collectively. We would like to specifically highlight the ongoing support and hard work of Carol Olivacz, Jack Hanley, and Kathy Josey, who've been tirelessly working with us for years, as well as all the former members of Rocky Mountain Paranormal.

Special thanks are due to Bret Smith and the entire family at the Colorado Festival of Horror. Their support has been immeasurable, not only as we prepared this book but as we've delivered our lectures and more over the years. And readers are encouraged to attend their annual festival and show them some love.

Additional thanks to Bob Bonner and Keith and Sheryl Lewis for ongoing support and to the late, great James "The Amazing" Randi for inspiration.

Thanks also to Aaron Bordner for the truly excellent interior artwork that helps bring some of these stories to life.

Finally, thanks to everyone who bought and read a copy of this book. Without you, none of this would mean anything.

About the Authors

Robert (Bob) Lewis is a Colorado-based author, editor, paranormal investigator, scholar, magician, and more. He holds degrees from the University of Colorado Denver in Biology, English, Mathematics, and Psychology, and a Master of Education from the University at Buffalo in Science and the Public. A dedicated polymath, he likes to tell people that his hobby is to collect new hobbies. He's (obviously) a member of the Rocky Mountain Paranormal Research Society. Additionally, he's a co-host of the Do You Like Scary Movies horror podcast, host of the Phobophile YouTube channel, and is always looking for more projects to whittle away at what little time he has left for sleep. He can be found online at www.robertlewisauthor.com.

Bryan Bonner is a founding member of the Rocky Mountain Paranormal Research Society. For over two decades, Bryan has dedicated himself to the examination of a wide range of paranormal phenomena, including ghosts, poltergeists, psychics, UFOs, conspiracy theories, urban legends, and much more. Setting himself apart from others in the field, Bryan has always maintained a grounded approach, refraining from running around cemeteries at night and needlessly scaring others with imaginative tales. Instead, he meticulously tests bizarre beliefs and practices, conducts experiments and on-site investigations, and even recreates unusual events, all with the aim of uncovering the truth. Bryan's work has garnered the respect of believers and skeptics alike, while simultaneously instilling fear in fraudsters and charlatans. You can read more about Bryan and his work at www.rockymountainparanormal.com.

www.ingramcontent.com/pod-product-compliance
Lightning Source LLC
Chambersburg PA
CBHW021609120626
46545CB00001B/134